澜湄职业教育培训中心暨柬埔寨鲁班工坊系列教材

A Series of Textbooks for Lancang-Mekong Vocational Education Training Center and Cambodia Luban Workshop

4G 通信网络管理员(高级)

4G Communication Network Administrator (Senior)

主　编　刘赟宇

Chief editor：LIU Yunyu

副主编　高　源　韩　健　高玉荣　孔　雷　张　磊

Deputy editors：GAO Yuan　HAN Jian　GAO Yurong

KONG Lei　ZHANG Lei

西安电子科技大学出版社

Introduction

Based on TD-LTE (Time Division-Long Term Evolution), this book adopts the modular teaching mode and applies the eNodeB equipment of Datang Mobile TD-LTE. Through the specific tasks in each module, it presents such contents as base station products, equipment installation, wireless technology, network testing, network planning and optimization, etc. in a comprehensive and detailed way. This book is divided into four training modules, namely, TD-LTE Equipment Cognition, TD-LTE Key Technologies, Network Test, Network Planning and Optimization. After learning the four training modules, learners can master the basic contents involved in the operation and maintenance of the mobile communications system, such as the hardware structure, maintenance and testing of the base station equipment, network planning and optimization. This book aims to cultivate high-quality skilled personnel who master the testing and optimization skills of base station and meet the requirements of standardized field maintenance operation.

On the basis of traditional technical resources and training materials, the common knowledge points about network planning and optimization are extracted, and systematic thinking is highlighted. With strong operability and practicality, the chapter layout of this book is based on the working process and practical tasks. Besides, this book explains the basic concepts of mobile communication in a simple way, comprehensively introduces the process of construction and maintenance, test and optimization of the 4G base station, and carries out network test and analysis with Datang Outum software.

This book can be used as the textbook of 4G communication technology courses for electronics, telecommunications and related majors of application-oriented universities and higher vocational colleges, as well as the reference book for mobile communications, network engineering professionals and engineering technicians.

图书在版编目（CIP）数据

4G 通信网络管理员：高级=4G Communication Network Administrator(Senior)：英文/刘赟宇主编. —西安：西安电子科技大学出版社，2022.2
ISBN 978-7-5606-6042-4

Ⅰ. ①4… Ⅱ. ①刘… Ⅲ. ①第四代移动通信系统—计算机网络管理—资格考试—自学参考资料—英文 Ⅵ. ①TN929.537

中国版本图书馆 CIP 数据核字(2021)第 249131 号

策划编辑　刘玉芳
责任编辑　王　斌　刘玉芳
出版发行　西安电子科技大学出版社(西安市太白南路 2 号)
电　　话　(029)88202421　88201467　　邮　　编　　710071
网　　址　www.xduph.com　　　　　　电子邮箱　xdupfxb001@163.com
经　　销　新华书店
印刷单位　咸阳华盛印务有限责任公司
版　　次　2022 年 2 月第 1 版　　2022 年 2 月第 1 次印刷
开　　本　787 毫米×1092 毫米　　1/16　印　张　6.5
字　　数　143 千字
印　　数　1～1000 册
定　　价　24.00 元
ISBN 978−7−5606−6042−4 / TN
XDUP 6344001-1
***** 如有印装问题可调换 *****

General Foreword

Serving the Belt and Road Initiative of China, the Lancang-Mekong Vocational Education Training Center and Cambodia Luban Workshop is a joint project undertaken by Tianjin Sino-German University of Applied Sciences(TSGUAS) for the Ministry of Foreign Affairs, the Ministry of Education and the Tianjin Municipal People's Government. Based in Cambodia, the project is designed to serve five countries in the Lancang-Mekong area and radiate to other ten ASEAN countries. It integrates functions of vocational training, vocational education, scientific research, cultural inheritance and innovation&entrepreneurship, develops both academic and non-academic education, and operates as a market-oriented international vocational training center.

At the initial stage of the project, 18 training rooms including mechanical processing technology, electrical technology and communication technology were built in three training centers for mechatronics and communication technology majors, with a total construction area of 6,814 m^2 and more than 1,600 sets of equipment.

The project will implement a "three-phase" plan. Based on the specialty construction in the first phase, international tourism, logistics engineering, automobile maintenance, building electricity and other specialties will be set up in the second phase to carry out technical skills training for Chinese&Cambodian enterprises and Cambodian people. Meanwhile, higher vocational education, applied technology undergraduate education, joint postgraduate education and other academic educations will be carried out to explore systematic talents cultivation of "medium and high vocational education, undergraduate education, and postgraduate education for a master's and doctoral degree".

Since 2017, as many as 95 articles about the project have been published by mainstream media including People's Daily, Guangming Daily, China Education News, Xinhuanet, etc. from home and abroad. After over two months of field study and research, Tianjin Television produced two feature stories named "Khmer Training", each lasting 30 minutes. The two episodes were broadcast on May 6th and May 13th 2019 respectively, featuring "on and on sails the vocational education, overseas shines the Luban Workshop". They give a full coverage of how TSGUAS teachers brought advanced skills to local areas and how friendship flourished along the Belt and Road Initiative route—a great contribution to the BRI. On July 18, 2019, the Royal Government of Cambodia conferred the Officer of the SAHAMETREI Medal to the Secretary of the Party Committee of TSGUAS, and the Knight of the SAHAMETREI Medal to the President and Vice President in charge of this project, with the signature of Prime Minister Hun Sen of Cambodia. On July 22, 2019, China Education Association for International Exchange awarded TSGUAS the medal of "Featured Cooperation Project of China-ASEAN Higher Vocational Colleges". In October 2019, the President of National Polytechnic Institute of

Cambodia (NPIC) presented 11 teachers with certificates and medals for their outstanding contributions to the Ministry of Labor and Vocational Training of Cambodia. Tianjin Sino-German University of Applied Sciences together with National Polytechnic Institute of Cambodia (NPIC) and their partners with enterprises was approved as the Belt and Road Joint Laboratory (Research Center)—Tianjin Sino-German and Cambodia Intelligent Motion Device and Communication Technology Promotion Center in December, 2020.

The Center has become a training base in Langcang-Mekong areas for technical talents training, a talent support base for Chinese enterprises overseas, a demonstration base for international students, and a base for teachers training. The Center is a key educational project of the Ministry of Foreign Affairs to serve the Belt and Road Initiative with foreign participation and entity institutions involved locally. The project will serve the social-economic and cultural development of the countries along the Initiative, enhancing the well-being of mankind; it will also serve the production output capacity of Chinese enterprises to help national development as well as enhance the international development of vocational education and the quality of its connotation. The project is a bridge connecting vocational education of Tianjin with the world, which marks a new stage of the city's international exchange and cooperation from a lower-medium to a medium-higher level.

The team of the project has compiled a series of textbooks for training, involving six occupations (electrotechnics, lathe, milling, CNC operation, bench and 4G communication network) from elementary, intermediate to advanced level based on current human resources situation in Langcang-Mekong countries, China's teaching equipment, and Chinese vocational qualification standards. These 19 textbooks target competence development and orient students to work tasks, combining theory with practice, and learning with practicing so as to put knowledge and skills into real situations. The textbooks aim to provide skills standards for the six occupations and lay foundations for the upgrading of the technological level of Lancang-Mekong countries.

ZHANG Xinghui

Party Secretary of Tianjin Sino-German University of Applied Sciences

June, 2021

A Series of Textbooks for Lancang-Mekong Vocational Education Training Center and Cambodia Luban Workshop Editorial Committee

Preface

This book is one of a series of tutorials for the "Lancang-Mekong Vocational Education and Training Center" project. The course is based on TD-LTE mobile communications standards and the research results of Tianjin Vocational Training Package (communication equipment inspector). Through school-enterprise cooperation, the training course is developed according to the skill requirements of specific job positions. It integrates skills training and vocational ability training and its content are compiled according to actual tasks of different job positions, from the network installation, network maintenance to network testing and network optimization of the 4th generation of Datang mobile communications system. Novel in lay-out and rich in content, the book is co-edited by a group of backbone teachers with rich teaching experience and senior engineers with rich engineering practice experience. In light of to the knowledge basis of trainees and the different levels of positions in enterprises, this set of teaching materials is divided into three volumes: 4G Communication Network Administrator - Primary (Level 5), 4G Communication Network Administrator - Intermediate (Level 4), and 4G Communication Network Administrator - Senior (Level 3).

Primary: Applicable for the new entry-level staff training, and those who need to acquire the basic engineering installation skills. Suitable posts are installation engineer, and supervision engineers.

Intermediate: Applicable for those who have mastered the basic engineering skills and have get two years of work experience in the project site and who are going to enter the intermediate work phase to learn about the skills of base station opening, debugging and operation and maintenance. Suitable posts are debugging engineers, and operation and maintenance engineers.

Senior: Applicable for those who have mastered intermediate engineering skills and have get one or two years of work experience in the project site and who are going to enter the senior work stage to study skills of base station testing, network planning and network optimization. Suitable posts are planning engineers and optimization engineers.

As an instruction for the senior administrators, on the basis of actual network construction of mobile communications network operators, this book comprehensively and systematically introduces the whole process of LTE wireless access network from installation and debugging, network planning, network testing to network optimization based on the mainstream product of the Datang Corporation, i.e. the EMB5116 base station. Besides, the book combines the theoretical basis of LTE with practical operation. The theoretical part focuses on the network structure, the wireless technology, and the important concepts and signaling process of LTE network while the practice part focuses on network testing and network planning and optimization. It builds a bridge between theory and practice for learners with certain knowledge basis about network communications, so that they can understand the process of LTE network testing and optimization with ease, and master the relevant skills of operation and maintenance.

It is available to scan the two-dimensional code below for corresponding contents of this book in Chinese.

Due to the limited knowledge of the author, any criticism and suggestions are welcome.

Managing Editor
2021.6

译文

Content

Contents

Training Module 1   TD-LTE Equipment Cognition

【Brief Description】

TD-LTE (Time Division Long Term Evolution) system consists of three major parts, base station (eNodeB, or eNB), evolved packet core network (EPC), and operation and maintenance center (OMC). The main task of mobile communication system installation is to recognize the devices of the three major parts, i.e. base station products, core network products, and OMC network management products and next master the installation skills of the equipment.

【Training Elements】

1. Knowledge objectives

(1) Master the technical features, appearance, and software and hardware structure of mobile communication system products.

(2) Master the service and functions of mobile communication system products.

(3) Master the equipment installation methods of mobile communication system products.

2. Capability objectives

(1) Master the basic composition of mobile communication system products.

(2) Master the service and functions of various mobile communication system products.

(3) Master the equipment installation methods of mobile communication system products.

【Training Requirements】

1. Preparation of tools, instruments and equipment

A set of mobile communication equipment and tools.

2. Knowledge appraisal point

(1) Composition of mobile communication equipment system.

(2) Functions of each component.

3. Skill assessment point

(1) Master the performance and function of main equipment.

(2) Master the installation specifications.

Task 1　Base Station

TD-LTE mobile communication network includes terminal, base station, core network, network management center, transmission network, etc., as shown in Fig. 1-1-1. Base station is the most important wireless device in the entire network.

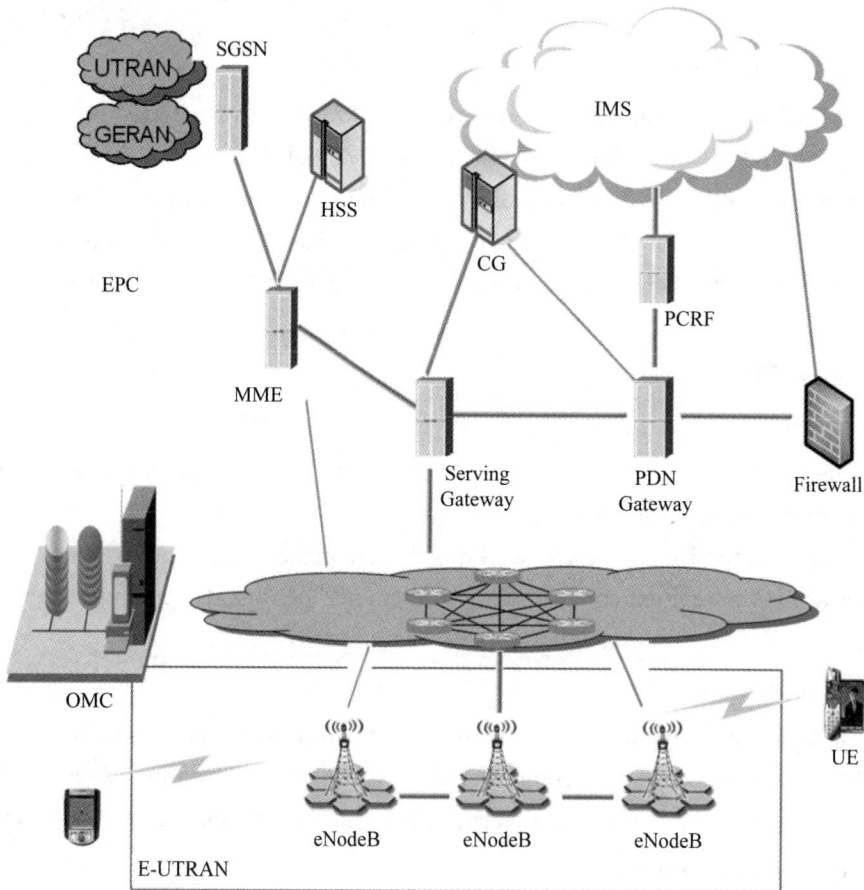

Fig. 1-1-1　TD-LTE network structure

1. Product Positioning

TD-LTE base station (indoor base station and outdoor base station) of Datang Mobile Communication Equipment Co., Ltd., can provide wireless access network solutions for TD-LTE system in response to different needs and actual situations of customers. EMB5116 is a baseband remote base station developed by Datang Mobile Communication Equipment Co., Ltd. Through the use of baseband remote technology, it can support both local coverage and remote coverage. It can be applied to outdoor macro coverage, such as urban hot spots, suburbs, towns, rural areas, or areas along the road, etc. By using the remote technology, the EMB5116 base station can quickly cover the main business areas at a low-cost. It can also be used to accomplish indoor coverage of small and medium capacity, such as tunnels, subway stations, buildings, residential areas, etc. The base station can improve network coverage and service quality without significantly increasing costs.

2. Technical Features

1) Wide coverage

The maximum local coverage radius is 100km; it supports smart antenna, which improves the uplink receiving sensitivity and increases the downlink coverage; the single-stage standard distance through optical fiber is 2km, the single-stage maximum pull-out capacity is 10km, the multi-stage pull-out is 40km at most, and the cascade level is 4 at most.

2) Flexible configuration

Each cell supports 400 active users and 1200 connected users with 20M bandwidth; standard configuration of three cells supports 60M bandwidth processing capacity at most, which can support 1200 active users and 3600 connected users; capacity expansion can be realized by adding baseband board; cell configurations of O1 and O2 are supported; cell configurations of S1/1/1 and S2/2/2 are supported.

3) Flexible networking

It supports networking of the same frequency, and supports S1/X2 port with star type, chain type and ring type networking, and Ir port with star type, chain type and ring type networking. Clock source supports GPS synchronization, Beidou satellite synchronization, GPS/Beidou satellite optical fiber pull-out synchronization, and superior eNodeB synchronization.

4) Flexible and convenient installation

It is small in size and light in weight, and can be easily installed on the indoor wall of the building. In the 19-inch standard cabinet, it can be directly installed in the indoor environment, with no need of machine room and air conditioning, so as to realize low-cost and rapid station construction.

5) Powerful operation and maintenance functions

The mobile network management system includes the operation and maintenance terminal such as LMT(Local Maintenance Terminal), providing management functions such as system status monitoring, data configuration, alarm processing, safety management, equipment operation, software configuration, monitoring management, self-configuration optimization, tracking management, etc.

6) Product seriation

The TD-LTE base station includes indoor base station and outdoor base station, which can provide wireless access network solutions for TD-LTE system in response to different needs and actual situations of customers.

3. Appearance of main equipment

The height of the EMB5116 TD-LTE case is 2U, and its overall dimension is 88mm ×

483mm × 310mm (H × W× D). Its appearance is shown in Fig. 1-1-2.

Fig. 1-1-2   Appearance of the EMB5116 TD-LTE case

4. Hardware structure

The EMB5116 TD-LTE main equipment includes switch control transmission board E type (SCTE), baseband processing only board X type (BPOX), common back plane (CBP), fan control Unit (FCU), environment monitor board A/D Type (EMA/EMD), power supply board A/C type (PSA/PSC), and extended transmission processing board E type (ETPE).

The layout of hardware units in the main unit is shown in Fig. 1-1-3. See Fig. 1-1-4 for full configuration.

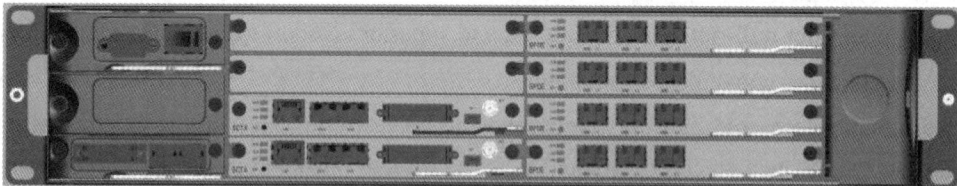

Fig. 1-1-3   Layout of hardware units in the main unit

SLOT 11	BPOG	SLOT 3	BPOG	SLOT 7	
PSC	BPOG	SLOT 2	BPOG	SLOT 6	FC
SLOT 10	SCTE	SLOT 1	BPOG	SLOT 5	SLOT 8
EMA SLOT 9		SLOT 0	BPOG	SLOT 4	

(a) Direct current

PSA SLOT 11	BPOG	SLOT 3	BPOG	SLOT 7	
	BPOG	SLOT 2	BPOG	SLOT 6	FC
SLOT 10	SCTE	SLOT 1	BPOG	SLOT 5	SLOT 8
EMA SLOT 9		SLOT 0	BPOG	SLOT 4	

(b) Alternating current

Fig. 1-1-4   Full configuration

1) Switching control and transmission unit

The switching control and transmission unit is composed of SCTE boards, whose main functions are as follows:

(1) The S1/X2 interface between EMB5116 TD-LTE and EPC can support 2 electric/optical ports, FE(Fast Ethernet)/GE(Gigabit Ethernet) adaptive interface types (2 interfaces can be configured as electric or optical ports respectively, supporting one optical one electrical) or their combination when the single box is fully configured.

(2) Service and signaling exchange function.

(3) All control and uplink interface protocol control surface processing.

(4) High stability clock and hold function.

(5) Power on and power saving control of single board card.

(6) In place detection and survival detection of single board card.

(7) Clock and synchronous stream distribution.

(8) Independent of the frame management of the board software.

(9) Redundant backup of the system.

2) Baseband processing unit

The baseband processing unit is composed of BPOG boards.

The main functions of BPOG board are as follows:

(1) Realize standard Ir interface.

(2) Realize the aggregation and distribution of baseband data.

(3) Implement TD-LTE physical layer algorithm.

(4) Realize L2 functions such as TD-LTE MAC / RLC / PDCP.

(5) Realize the operation and maintenance of the board.

3) Environmental monitoring unit

The environmental monitoring unit is responsible for environmental monitoring.

4) Fan unit

The fan unit is responsible for cooling the equipment.

5) Power supply unit

The power supply unit is responsible for supplying the equipment with power.

6) Extended transmission processing unit

The extended transmission processing unit is responsible for the extended transmission function of the device.

5. RF unit appearance

The base station RF unit includes two-channel RRU and eight-channel RRU.

1) Two-channel RRU appearance

The two-channel RRU includes TDRU331FAE, TDRU332FA, TDRU341FAE, TDRU342D and TDRU342E.

　　TDRU342E dimension: 420mm × 300mm × 120mm (H × W × D), net weight of case: 12kg, equipment capacity: 15L. Its appearance is shown in Fig. 1-1-5.

　　2) Eight-channel RRU appearance.

　　The eight-channel RRU includes TDRU338FA, TDRU338D and TDRU348FA.

　　TDRU338D dimension: 439mm × 356mm × 140mm (H × W × D), net weight of case: 23kg, equipment capacity: 23L. Its appearance is shown in Fig. 1-1-6.

Fig. 1-1-5　TDRU342E

Fig. 1-1-6　TDRU338D

Task 2　Core Network

　　LTE core network provides users with IP transmission services with large data bandwidth, low service delay, fast access speed and strong security. It supports a variety of different access technologies, enabling users to obtain uninterrupted services when moving between different access networks. It supports emergency call and roaming user routing optimization. It provides operators with flexible control strategies and accounting methods, as well as abundant operation and maintenance methods for different products.

1. Definition of core network

　　EPC core network is composed of MME（Mobility Management Entity), S-GW(Serving Gateway), P-GW (PDN Gateway), PCRF (Policy and Charging Rules Function), CG (Charging Gateway), HSS (Home Subscriber Server), etc.

2. MME

　　The functions of MME are mainly access authentication, mobility management, session management, switch control and clock synchronization.

3. S-GW

　　The functions of S-GW are mainly carrying management, route selection and data forwarding, QoS control, GTP and accounting.

4. P-GW

　　The functions of P-GW are mainly UE IP address management, bearer management, routing and data forwarding, security, accounting, etc.

5. PCRF

Policy Based Charge Control (PCC) technology is that when the user's service data flow needs to be transmitted through the PS domain, the network element PCRF performs dynamic QoS and charge policy control of the user's application level service data flow in view of the characteristics of the service data flow, the operator's strategy and the subscription characteristics of the subscribers, so as to realize the operators effective control and management of the user's service.

6. CG

The accounting gateway (CG) is an important network element in TD-LTE EPC core network. Its main function is to obtain the original bill record from S-GW and P-GW, and then to process the bill to form the final bill file. The final bill file is transferred to BOSS(Business Operation Support System) by FTP mechanism for processing.

7. HSS

Home Subscriber Server is an important network element in LTE EPC, which contains user profile, performs user authentication and authorization, and provides information about user's physical location. HSS can complete user information subscription user authentication and other functions in LTE.

Task 3   OMC Network Management System

1. Product positioning

OMC network management system provides configuration, alarm, performance, security, log, software and other management work for access network equipment, as shown in Fig. 1-3-1.

OMC network management system is responsible for the maintenance of telecommunication network equipment, providing man-machine interface, configuration, inspection, control, diagnosis and operation authority inspection of telecommunication network equipment, tracking the operation status of equipment, collecting and analyzing the operation data of network equipment.

OMC network management system provides the management and maintenance of each network element of wireless equipment eNodeB, core network equipment EPC and IMS (IP Multimedia Subsystem), and also of RNC and eNodeB which are compatible with 3G access network.

OMC network management system provides real-time monitoring service for network element equipment and network, supports simple and quick data configuration, software upgrade and system expansion, and achieves complete and comprehensive performance index counting, all of which can offer a strong support for convenient and efficient work of various users. OMC network management system personnel are classified as follows:

(1) Network optimization personnel: collect and optimize network data through OMC

network management system.

(2) Project maintenance personnel: through OMC network management system, undertake major tasks including the monitoring, maintenance and data configuration of network equipment, network element equipment opening, fault location and troubleshooting, network expansion, centralized upgrading, etc.

(3) High level network management: accomplish network level operation and maintenance management of alarm log, performance index, and parameter configuration through OMC network management system.

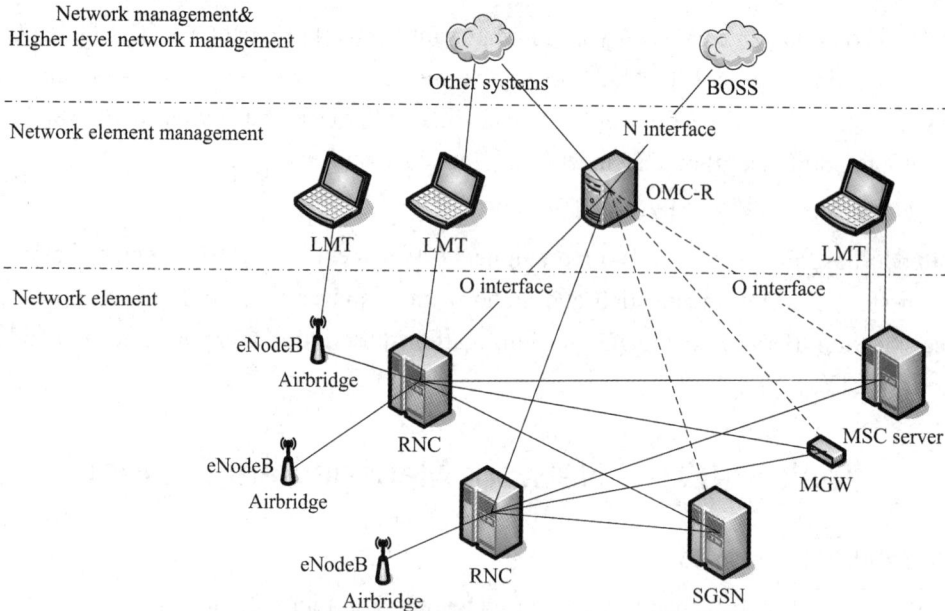

Fig. 1-3-1   OMC network management system

2. Technical features of products

1) Centralized network management and large capacity

OMC is mainly used for the centralized management of multiple devices to monitor the operation of the entire network. Therefore, its multi transaction processing capability and performance counting function are more prominent than those of LMT (Local Maintenance Terminal), and it is mostly used in the process of network operation, maintenance and optimization.

Besides the function of centralized network management, the OMC network management system of Datang Mobile also integrates some functions of LMT, which not only meets the requirements of centralized network management personnel, but also provides support for the application of special scenarios.

2) Multiple functions and convenient operation

The OMC network management system supports the centralized management of eNodeB, EPC and IMS network elements of all models of Datang Mobile. Through the graphical interface, it can complete such functions as alarm management, configuration and status management performance statistics, security and log management, software management, topology management, system management, test diagnosis, and MML.

3) Fault tolerance

(1) Unidentifiable messages reported by the device will be discarded directly without affecting the subsequent message processing.

(2) For the content that cannot be analyzed in the performance data file, the system will try its best to ensure effective data warehousing.

(3) For the query of the properties of configuration object, if the value is out of range, its original value will be displayed and marked with a bright color.

(4) During the execution of the command line script, if the execution of a single command fails, the system can abort the execution of the command, prompt for errors or skip directly.

(5) For irregularities in the process of program processing, the system will capture and record the log normally, which only affects the current transaction operation.

4) Maintainability

(1) OMC provides log hierarchical recording function, which records system operation exceptions, key process stage points, and message interaction between external systems.

(2) OMC provides one click operation to get the operation log.

(3) OMC supports remote viewing of system resource usage.

(4) OMC supports patch loading and upgrading without the need to restart the process.

The designs for maintainability-related functions include platform log recording, log query and audit, system process management, and OMC software upgrade.

5) Recoverability

If the server process exits under abnormal conditions, it can be actively pulled up by the system monitoring process. When the communication with the external system (including the network element and the superior network management) is interrupted, it can be restored automatically. For functions relevant to recoverability please refer to system process management and platform communication layer design.

6) Easy installation

For different operating systems, it is necessary to provide professional software installation wizard tools, such as Installshield tool and WebStart technology, to simply and quickly complete the software installation and deployment, and to provide detailed and operable software installation instructions. The features of easy installation are listed in the following :

(1) Client software can either be installed or operated on demand.

(2) When the software is upgraded, differences between different versions are automatically compared, and the updated incremental patch and version are downloaded.

(3) The Installshield tool is used to simplify the installation process of the OMC software. For the installation of the client, WebStart technology is adopted to address the need of upgrading the client at the same time.

(4) The client software is divided into user management, data configuration, alarm monitoring, performance data management and other professional tools, which can be installed and started on demand.

7) Adaptability

OMC network management system can support a variety of mainstream hardwares, operating systems and databases. Using J2EE technology, it can generate different hardware platform installation packages through Installshield. For the implementation of differential functions, such as system monitoring and FTP access, they need to be implemented on the hardware platform.

8) Multi-language

OMC network management system supports multi-language display. Language and time zone can be automatically selected according to the regional configuration of the operating system. At least simplified Chinese and English are supported. Any content to be displayed needs to be implemented in accordance with the international coding specification provided by the platform.

9) Support OEM

By simply replacing configuration files, pictures and other contents, OEM (Original Entrusted Manufacture) products with the same baseline can be constructed. All the resources on display are obtained by the resource acquisition method defined by the platform, and different resource packages are encapsulated by the resource acquisition method (corresponding to different manufacturers).

10) Testability

The testability of software is the ability to detect software problems, isolate and locate faults by testing design and execution at a certain time and cost. Simply put, the testability of software is the ease with which a computer program can be tested.

The testability features of software mainly include setting up observation point, control point, observation device, driving device and isolation device. It should be noted that testability design must ensure that it cannot affect any function of the software system, and cannot generate additional activities or additional tests, and that on appropriate design mode is adopted to design the software.

Testability is guaranteed by the following steps:

(1) Object oriented design and Java development.

(2) Separation of data display and control.

(3) Modular design.

(4) Interface oriented design.

(5) Aspect oriented design (realized through OP(Operation Platform)).

(6) Configurable design of service process (implemented by introducing workflow engine).

(7) Standardization of log output.

11) Capacity demand

(1) Network element management capabilities. OMC network management system supports devices that can manage core network and access network at the same time. with the standard configuration, it can meet the management needs of all devices in large and medium cities.

(2) OMC management capabilities. They can be improved by increasing the number of service servers and network element adapters.

(3) Data storage capabilities. OMC network management system can save at least 3 months of performance data, 3 months of alarm data, 12 months of user operation logs, and 3 years of performance reports.

When designing the OMC configuration list, it is necessary to convert the required data storage space according to the network scale and service flow, plus 20% of the configuration margin.

12) Performance requirements

The data synchronization time of a single NE should be less than or equal to 1min. As the key technical indicators, they should be met at the configuration management outline design stage.

(1) Alarm receiving capability.

(2) Ensure orderly and reliable processing.

(3) Guarantee the processing capacity of 300 per second.

(4) The delay from the report of NE to the display on the terminal does not exceed 5 s.

13) Safety

(1) Operation confirmation function. Important changes and deletions must be confirmed by the user before the command is issued. For operations that have a greater impact on the system, such as resetting, the consequences of the operations should be displayed in a prompt dialog box. For security management, it can be configured whether the operation needs to be confirmed and whether password verification is required.

(2) Message encryption function. In order to prevent hackers from attacking network element devices, OMC and network element interface messages should be encrypted. More general algorithms are adopted as encryption algorithms, such as SSL (Secure Socket Layer), MD5 and so on.

① Between OMC and network elements, secure channel technology is used, such as the SNMPV3 protocol for user authentication.

② OMC internal communication can be configured to use SSL interface encryption.

③ The northbound interface can choose whether to use SSL encryption according to the specifications.

(3) Detection of abnormal process and port. It supports the configuration of process and port trust list, and provides alarm prompt for start process and use port out of the list. It is designed and implemented as a system management function.

14) Openness

OMC network management system supports the required data opening A configuration, performance, alarm and log data, and supports all-round data sharing at database level and file level. OMC can export general data to file (CSV, xls, XML and other types), supporting database view sharing configuration and sharing data on demand.

Task 4　Base Station Installation

This task takes Datang Mobile's EMB5116 TD-LTE as an example. It is a compact base station product developed to adapt to a variety of possible application environments. It is small in size, simple in installation and flexible in configuration. It can be fixed on walls or inside cabinets, etc. It can be used for indoor distributed applications and outdoor macro cellular applications. With a full compatibility design, it can provide operators with wide networking applications together with other TD-LTE base station products of Datang Mobile. This project is a typical work task designed for the practitioners who are involved in the network installation of mobile communication system for the first time. The task mainly takes two typical base station installation methods as the training objectives. The contents include base station system structure, installation preparation, main equipment installation, RF unit installation, GPS system installation, lightning protection equipment installation, grounding kit installation, etc.

The connection of base station system(outdoor) is shown in Fig. 1-4-1.

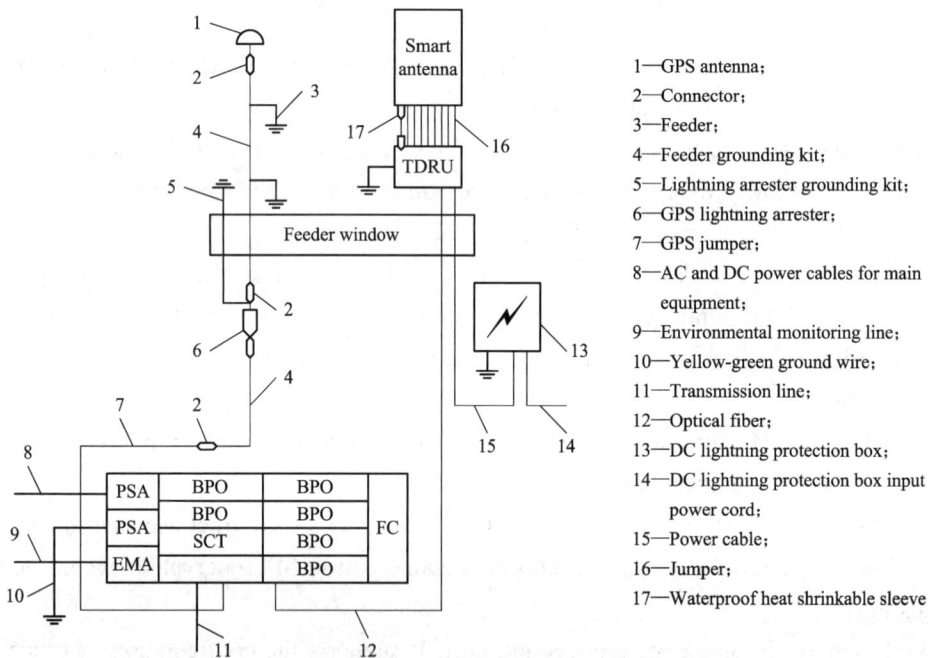

1—GPS antenna;
2—Connector;
3—Feeder;
4—Feeder grounding kit;
5—Lightning arrester grounding kit;
6—GPS lightning arrester;
7—GPS jumper;
8—AC and DC power cables for main equipment;
9—Environmental monitoring line;
10—Yellow-green ground wire;
11—Transmission line;
12—Optical fiber;
13—DC lightning protection box;
14—DC lightning protection box input power cord;
15—Power cable;
16—Jumper;
17—Waterproof heat shrinkable sleeve

Fig. 1-4-1　Connection of base station system(outdoor)

1. Preparations

1) Installation environment inspection and verification

To ensure the safe, stable and reliable operation of the EMB5116 TD-LTE compact base station, the first thing to be considered is to set the base station in a good working environment, rather than in an area with high temperature, dust, harmful gas and inflammable and explosive substances, at places with strong vibration and big noise, or at locations with the substation and high-voltage transmission line. The house structure, heating and ventilation, power supply and water supply, lighting and fire control in the equipment room should be designed and constructed in accordance with the relevant national and industrial standards.

   2) Preparation of installation tools

The tools required for installation are shown in Table 1-4-1.

Table 1-4-1   Installation tools

Tools	Measuring and marking tools: long tape, level, marking pen
	Drilling tools: percussion drill, matching drills, vacuum cleaner
	Fastening tools: slotted screwdriver, cross screwdriver, medium spanner, socket wrench, ring wrench, hexagonal wrench
	Fitter tools: pointed nose pliers, diagonal pliers, vises, files, hand saws, wire strippers, handle crimping pliers, wire cutters, RJ-45 crystal head crimping pliers, hydraulic pliers
	Auxiliary tools: brush, medium claw hammer, paper cutter, blower, electric soldering iron, solder wire, ladder, rubber hammer, compass, torque wrench, chamfered, hot air gun (electric hot air gun or liquefied gas hot air gun)
	Antistatic wrist strap and gloves
Meters	Multimeter, 500V megohmmeter (for insulation resistance measurement), optical power meter, ground resistance meter, antenna standing wave ratio tester, inclinometer

2. Tasks

Base station installation.

Skill 1   Main Equipment Installation

Three ways are used to install the main equipment (EMB5116 TD-LTE) case: inside a 19-inch cabinet, fixed on wall, use supporting frame/chassis.

   1) Inside a 19-inch cabinet

Requirements of a 19-inch standard cabinet for EMB5116 TD-LTE installation: it is required to provide an installation space of 2U high and more than 500mm deep, and the wiring space of 19-inch cabinet rack column should be 100mm from the front door. The cabinet meeting the above requirements can be used to install the EMB5116 TD-LTE.

   2) Wall-mounted installation

When the main equipment of EMB5116 TD-LTE is hung on the wall, the wall should be cement wall or brick (non hollow brick) wall, and the wall's thickness should be greater than 70mm.

   3) Supporting frame/chassis

When the EMB5116 TD-LTE frame is installed, the wall should be cement wall or brick (non hollow brick) wall, and the wall's thickness should be greater than 70mm. The installation steps of the frame in three directions are the same.

4) Installation of main equipment cable

The grounding wire, power line, NB-RRU optical fiber, transmission line and 26 pin environmental monitoring line of EMB5116 TD-LTE are all in the form of front panel outgoing line. The EMB5116 TD-LTE needs to be connected with the following cables: -48V DC power line or 220V AC power line, NB-RRU optical fiber, transmission line, grounding wire and 26 pin environmental monitoring line.

The connection position of each cable is shown in Fig. 1-4-2.

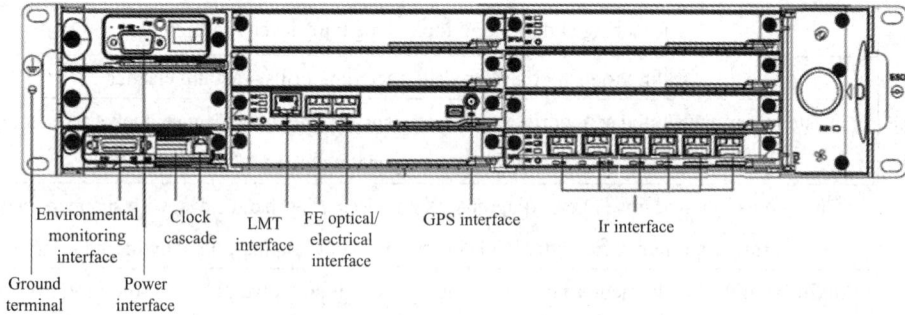

Fig. 1-4-2   EMB5116 TD-LTE panel interface

(1) Installation of main equipment grounding wire. RVVZ single core (16mm^2) (yellow and green) wire is used as the main equipment grounding wire, which can be cut according to the actual length. One end is connected to the main equipment grounding terminal, and the other end is connected to the indoor grounding bar, as shown in Fig. 1-4-3.

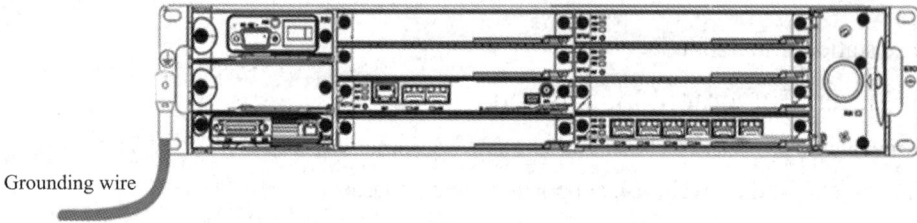

Fig. 1-4-3   Installation of main equipment grounding wire

(2) Installation of main equipment power line, as shown in Fig. 1-4-4.

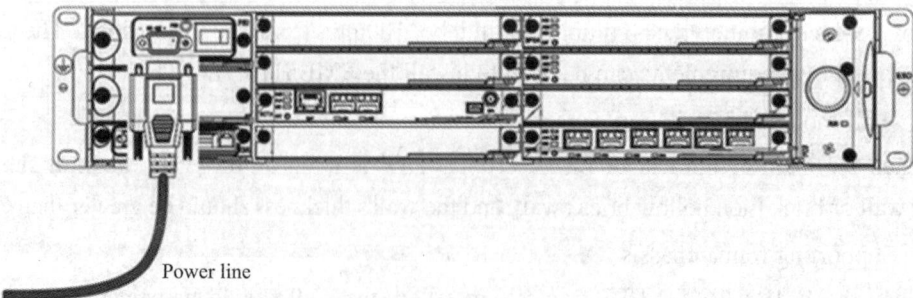

Fig. 1-4-4   Installation of main equipment power line

(3) Installation of NB-RRU optical fiber. According to the design requirement, connect the NB-RRU optical fiber linked with the RRU to the corresponding optical module on the front panel of the BPOG board card of the main equipment, as shown in Fig. 1-4-5.

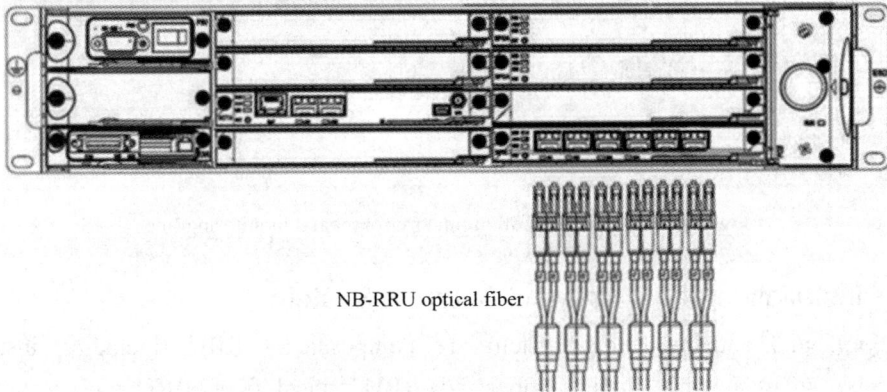

NB-RRU optical fiber

Fig. 1-4-5   Installation of NB-RRU optical fiber

(4) Installation of transmission line. In the EMB5116 TD-LTE, the S1/X2 interface supports GE/FE, and the external interfaces are on the left side of the SCTE board, which are two RJ-45 interfaces or optical interfaces, as shown in Fig. 1-4-6.

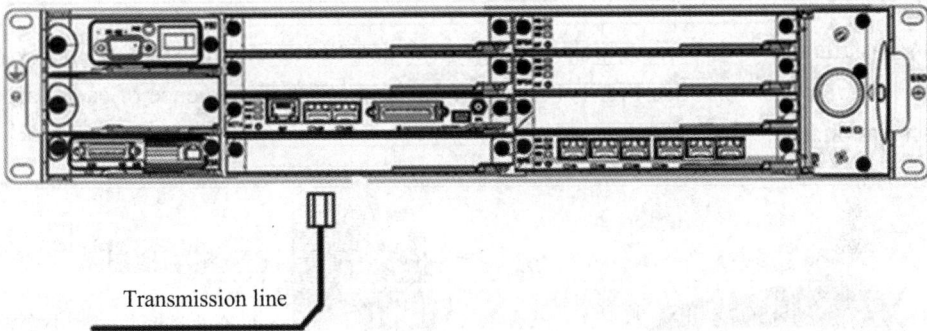

Transmission line

Fig. 1-4-6   Installation of transmission line

The external basic configuration of EMB5116 TD-LTE supports adaptive Ethernet, and the basic configuration is in the form of SFP optical interface. When electric connection is needed, the system can realize the connection by selecting RJ-45 to SFP switching accessories.

One end is installed on the SCTE panel and the other end is installed on the optical interface board of ODF. Each SCTE is connected with the optical transceiver through DLC single-mode optical fiber.

(5) Installation of 26 pin environmental monitoring line. The environmental monitoring of EMB5116 TD-LTE is implemented in the EMA unit. When the environmental monitoring is implemented through the dry node mode, the environmental monitoring cable with SCSI 26 pin at one end is required to be connected to the environmental monitoring port of the main equipment, as shown in Fig. 1-4-7.

26 pin environmental
monitoring line

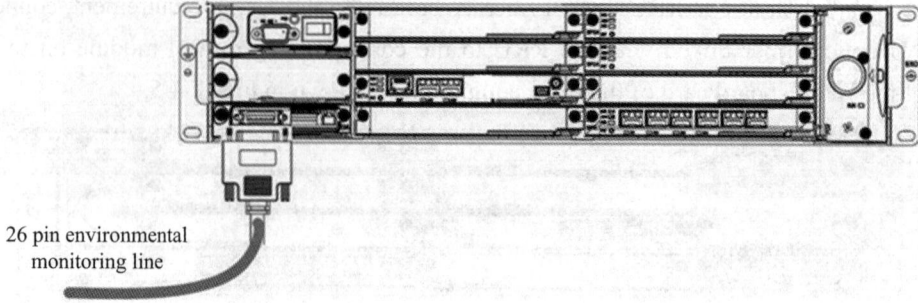

Fig. 1-4-7   Installation of 26 pin environmental monitoring line

Skill 2   Installation of RF Unit and Antenna System

RF unit and antenna feeder system are composed of RRU (including installation components), smart antenna, upper jumper, NB-RRU optical fiber, RRU power cord, RRU grounding kit, etc.

1) Installation of RF unit

The radio frequency unit (TDRU338D) is shown in Fig. 1-4-8.

There are two ways to install RRU chassis: installation with holding pole and installation on walls.

(1) Installation with holding pole. The general installation fixture supports poles with a diameter of 50mm-114mm. See Fig. 1-4-9 for the RRU installation sequence of each part of the pole installation.

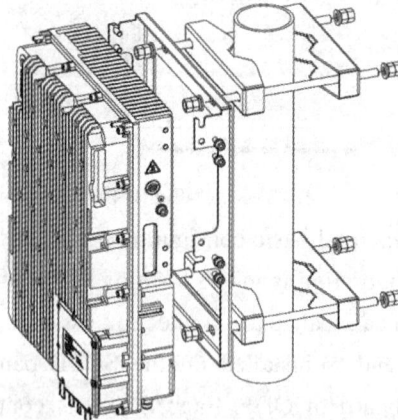

Fig. 1-4-8   TDRU338D   Fig. 1-4-9   RRU installation sequence of each part of the pole installation

(2) Installation on walls, as shown in Fig. 1-4-10.

2) Installation of antenna

The antennas used with TDRU338D are 8-element dual polarization sector smart antenna and 8-element omnidirectional smart antenna (ring array). See Fig. 1-4-11. for the installation of 8-element dual polarization sector smart antenna.

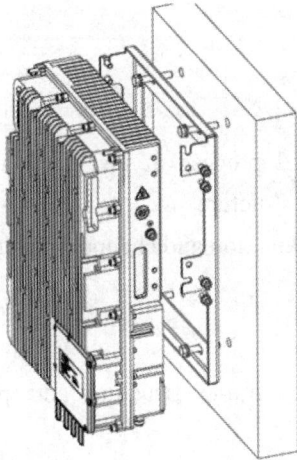

Fig. 1-4-10   RRU installation sequence of each part   Fig. 1-4-11   Installation of 8-element dual polarization

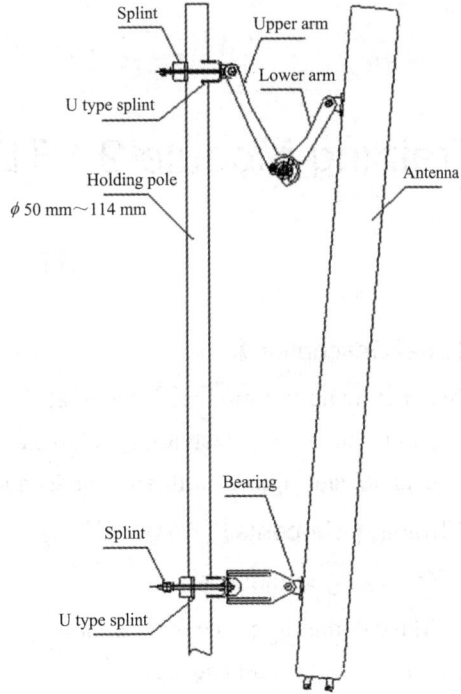

of the wall installation   sector smart antenna

Skill 3   Installation of GPS system

GPS system consists of GPS antenna (including antenna installation components), GPS feeder, GPS arrester, GPS jumper, GPS lower jumper, SCTE board (located in BBU), GPS feeder grounding kit, GPS arrester grounding kit, etc. The standard structure of the GPS system is shown in Fig. 1-4-12.

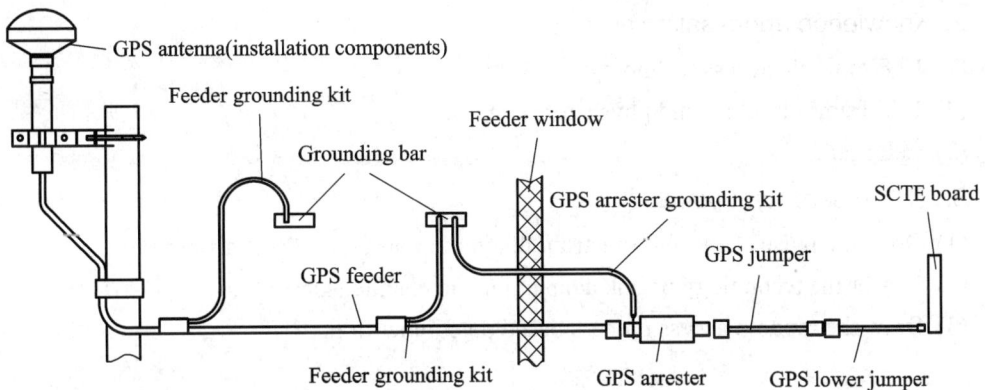

Fig. 1-4-12   Standard structure of the GPS system

Make a return water bend, tighten the fasteners and seal the feeder window before entering the feeder into the feeder window. Besides, a return water bend should be made to prevent rainwater from seeping into the indoor main equipment along the feeder.

Training Module 2　TD-LTE Key Technology

【Brief Description】

This part mainly introduces the related technology and process of the physical layer, including multiple access technology, duplex mode, frame structure and physical resources, physical channel and signal, multi-antenna technology, link adaptation and channel scheduling.

【Training Elements】

1. Knowledge objectives

(1) Master multiple access technology, duplex mode, frame structure and physical resources, physical channel and signal.

(2) Understand multi-antenna technology, link adaptation and channel scheduling.

2. Capacity objectives

(1) Analyze the relationship between wireless technology and network characteristics.

(2) Use wireless technology to solve network fault and optimize network.

【Training Requirements】

1. Preparation of tools, instruments and equipment

A set of mobile communication equipment.

2. Knowledge appraisal point

(1) LTE multiple access technology.

(2) LTE frame structure and physical resources.

(3) Cell search.

3. Skill assessment point

(1) Draw the pulse shape and spectrum diagram of an OFDM subcarrier.

(2) Master the technology of link adaptation and channel scheduling.

(3) Draw the random access process based on competition.

Task 1　LTE Interface

1. Wireless interface protocol

Wireless interface is the interface between terminal and eNodeB (Evolved Node B), which is a completely open interface. The wireless interface protocol stack is mainly divided into three layers and two planes, the three layers include the physical layer, data link layer and network layer, the two planes of wireless interface include the control plane and user plane, as shown in Fig. 2-1-1.

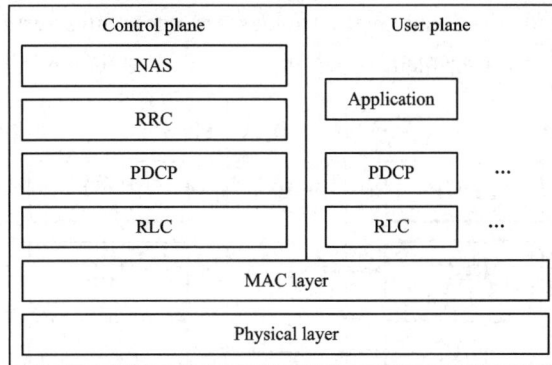

Fig. 2-1-1   Wireless interface protocol stack

Below are the three layers of the wireless interface protocol stack:

(1) First layer: Physical Layer. The physical layer provides wireless resources for the high level layer data and processes the physical layer.

(2) Second layer: MAC(Media Access Control)-PDCP(Packet Data Convergence Protocol), distinguishes and identifies data and provides service.

(3) Third layer: RRC(Radio Resource Control), the user of air interface service, i.e. RRC signaling and user interface service data.

Below are the two planes of the wireless interface protocol stack:

(1) The control plane of wireless interface is mainly responsible for the management and control of wireless interface, including RRC protocol, data link layer protocol and physical layer protocol. The data link layer is divided into three sub layers, media access control, wireless link control and packet data convergence protocol. NAS (Non-Access Stratum) control protocol entity is located in UE (User Equipment) and MME (Mobility Management Entity) and is responsible for the control and management of non-access layer parts. eNodeB does not handle NAS. RRC protocol entities are located in UE and eNodeB network entities and are responsible for the control and management of access layer. Data link layer and physical layer provide data transmission function for RRC protocol messages. LTE control plane of wireless interface is shown in Fig. 2-1-2.

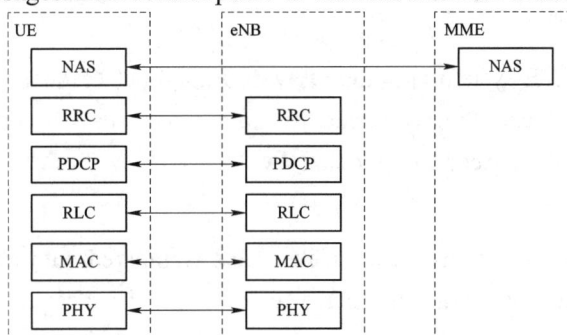

Fig. 2-1-2   LTE control plane of wireless interface

(2) The user plane protocol of wireless interface includes data link layer protocol (MAC, RLC, and PDCP) and physical layer protocol. The physical layer provides data transmission for the data link layer. Physical layer provides corresponding services for MAC sublayer through transmission channel, and MAC sublayer provides corresponding services for RLC sublayer through logical channel. LTE user plane of wireless interface is shown in Fig. 2-1-3.

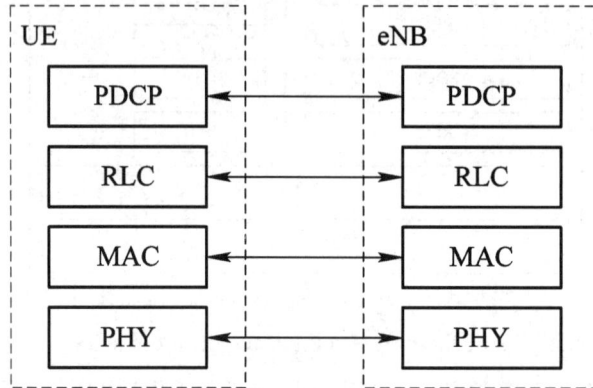

Fig. 2-1-3   LTE user plane of wireless interface

2. S1 Interface protocol

The control plane of S1 interface provides the function of control plane between eNodeB and MME. The control plane includes an application protocol and a signaling bearer for transmitting application protocol messages. The user plane of S1 interface provides the function of user data transmission between eNodeB and S-GW (Service GateWay). The user plane includes data bearer for data flow, which refers to the tunnel protocol of transmission network layer.

3. X2 Interface protocol

X2 interface is the interface between eNodeB and eNodeB. The definition of X2 interface adopts the same principle as S1 interface. The control plane protocol structure and user plane protocol structure of X2 interface are similar to those of S1 interface.

Task 2   LTE Multiple Access Technology

The downlink of LTE system uses the OFDM(Orthogonal Frequency Division Multiplex) technology with CP (Cyclic Prefix), while its uplink uses the SC-FDMA (Single Carrier Frequency Division Multiple Access) technology with CP.

1. OFDM

OFDM is a multi-carrier transmission technology. Compared with traditional multi-carrier transmission, OFDM can use more and narrower orthogonal subcarriers for transmission, as shown in Fig. 2-2-1.

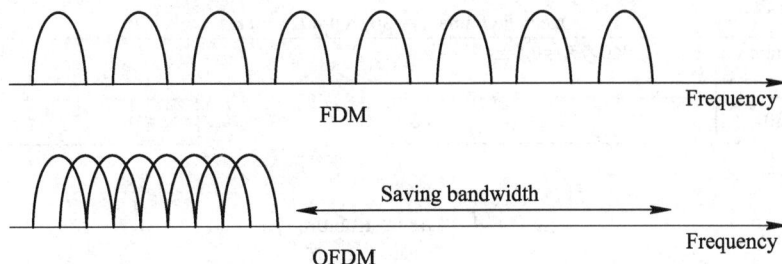

Fig. 2-2-1   OFDM vs. FDM

An OFDM symbol is composed of several such subcarriers, which are overlapped and orthogonal to each other in the frequency domain. Δf is the subcarrier interval, as shown in Fig. 2-2-2.

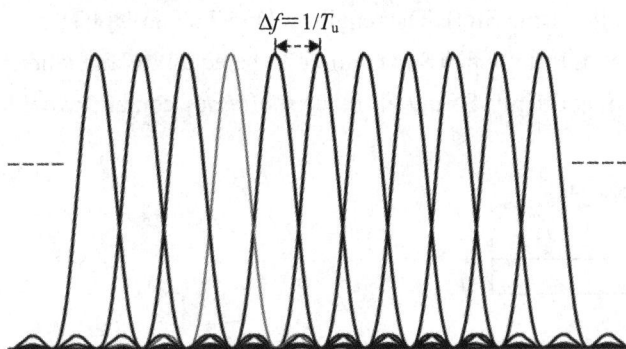

Fig. 2-2-2   Spectrum diagram of an OFDM symbol

The orthogonality of OFDM means that any two subcarriers in an OFDM symbol are orthogonal to each other. Because of this orthogonality, the demodulation of OFDM is relatively simple.

Task 3   LTE Frame Structure and Physical Resources

1. Frame structure

LTE supports two frame structures, which are suitable for FDD (Frequency Division Duplex), H-FDD and TDD (Time Division Duplex) operations. LTE defines the basic time unit as $T_s - 1/(15000 \times 2048)$ second.

1) Frame structure type 1

Frame structure type 1 is applicable to FDD mode of full duplex and half duplex. The length of each wireless frame is 10ms, composed of 20 time slots, and the length of each time slot is 0.5ms. These time slots are numbered 0 to19. A subframe is defined as two adjacent time slots, as shown in Fig. 2-3-1.

For FDD, in each 10ms, 10 subframes can be used for downlink transmission and 10 subframes can be used for uplink transmission. Uplink and downlink transmissions are separated in frequency domain.

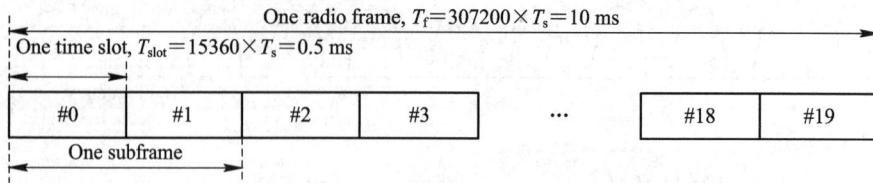

Fig. 2-3-1   Frame structure type 1

2) Frame structure type 2

Frame structure type 2 applies to TDD mode. Each wireless frame is composed of two half-frames, each half-frame being 5ms long. Each half-frame includes 8 time slots with a length of 0.5ms and three special time slots, DwPTS (Downlink Pilot Time Slot), GP (Guard Period) and UpPTS (Uplink Pilot Time Slot). The length of DwPTS and UpPTS is configurable, and the total length of DwPTS, GP and UpPTS is required to be equal to 1ms. Subframe 1 and subframe 6 contain DwPTS, GP and UpPTS, and all other subframes contain two adjacent time slots, as shown in Fig. 2-3-2.

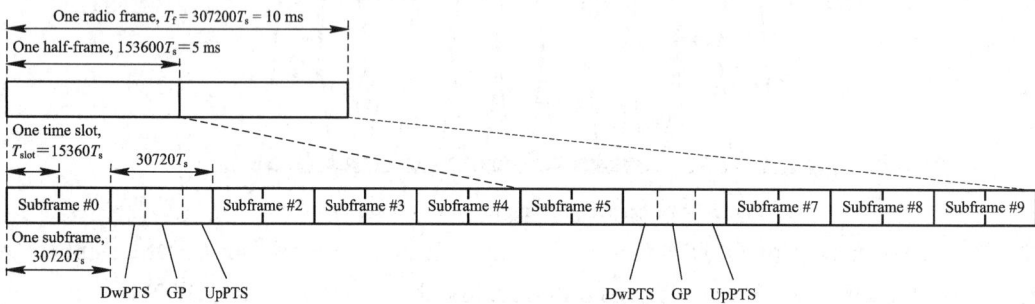

Fig. 2-3-2   Frame structure type 2

Frame structure type 2 supports up and down subframe switching cycles of 5ms and 10ms, and the specific configuration is shown in Table 2-3-1.

Table 2-3-1   Configuration of frame structure type 2

Configuration	Switch period	Subframe number									
		0	1	2	3	4	5	6	7	8	9
0	5 ms	D	S	U	U	U	D	S	U	U	U
1	5 ms	D	S	U	U	D	D	S	U	U	D
2	5 ms	D	S	U	D	D	D	S	U	D	D
3	10 ms	D	S	U	U	U	D	D	D	D	D
4	10 ms	D	S	U	U	D	D	D	D	D	D
5	10 ms	D	S	U	D	D	D	D	D	D	D
6	10 ms	D	S	U	U	U	D	S	U	U	D

The special time slot configurations of three special subframes are shown in Table 2-3-2.

Table 2-3-2   Configuration of special subframes

configuration	Regular CP			Extended CP		
	DwPTS	GP	UpPTS	DwPTS	GP	UpPTS
0	3	10	1	3	8	1
1	9	4	1	8	3	1
2	10	3	1	9	2	1
3	11	2	1	10	1	1
4	12	1	1	3	7	2
5	3	9	2	8	2	2
6	9	3	2	9	1	2
7	10	2	2	—	—	—
8	11	1	2	—	—	—

2. Physical resources

The smallest resource unit used in uplink and downlink transmission is called resource element(RE). The concept of resource block(RB) is defined based on RE. One RB contains several REs.

1) Resource particles

A unit corresponding to a subcarrier on an OFDM or SC-FDMA symbol is called a resource unit, as shown in Fig. 2-3-3.

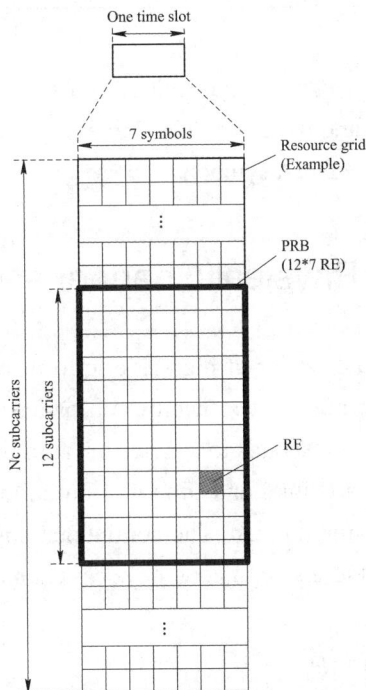

Fig. 2-3-3   RE, PRB and resource grid

(taking subcarrier spacing of 15 kHz and conventional CP as examples)

The number of symbols contained in each time slot depends on the CP used. A regular CP time slot contains 7 symbols, and an extended CP time slot contains 6 symbols, as shown in Table 2-3-3.

Table 2-3-3   Number of symbols in time slots under different CP

Subcarrier spacing	CP	OFDM / SC-FDMA number of symbols
$\Delta f = 15$ kHz	Regular CP	7
	Extended CP	6
$\Delta f = 7.5$ kHz	Extended CP	3

2) Physical resource block

In a time slot, the physical resources with a continuous width of 180kHz in the frequency domain are called a physical resource block (PRB), as shown in Fig. 2-3-3.

Represented by the number of subcarriers and symbols, the relationship between PRB and RE is shown in Table 2-3-4.

Table 2-3-4   PRB size

Subcarrier spacing	CP	Number of subcarriers	OFDM/SC-FDMA symbol number	RE number
$\Delta f = 15$ kHz	Regular CP	12	7	84
	Extended CP	12	6	84
$\Delta f = 7.5$ kHz	Extended CP	24	3	72

3) Resource grid

All the resource units occupied by the transmitted signal in a time slot form a resource grid, which contains an integer PRB, and which can also be represented by the number of subcarriers and the number of OFDM or SC-FDMA symbols.

Task 4   Physical Channel and Signal

Physical channel is the actual carrier of high level information in wireless environment. In LTE, the physical channel is determined by a specific subcarrier, time slot and antenna port. That is, on a specific antenna port, it corresponds to a series of radio time-frequency resource element (RE). A physical channel has start time, end time and duration. The physical channel can be continuous or discontinuous in time domain. The continuous physical channel duration is from the beginning to the end, while the discontinuous physical channel must be clearly indicated by of which time slices it is made.

1. Downlink physical channel

The downlink physical channel includes:

(1) Physical Downlink Shared Channel, PDSCH.

(2) Physical Multicast Channel, PMCH.

(3) Physical Downlink Control Channel, PDCCH.

(4) Physical Broadcast Channel, PBCH.

(5) Physical Control Format Indicator Channel, PCFICH.

(6) Physical Hybrid ARQ Indicator Channel, PHICH.

2. Downlink physical signal

The downlink physical signal includes the downlink reference signal and the downlink synchronization signal.

(1) The downlink reference signal is essentially a pseudo-random sequence, which does not contain any actual information. This random sequence is sent out through the resource unit RE composed of time and frequency, which is convenient for the receiver to estimate the channel and provides reference for the receiver to demodulate the signal. The downlink reference signal includes:

① Cell specific reference signal, associated with non-MBSFN transmission.

② MBSFN reference signal, associated with MBSFN transmission.

③ Terminal specific reference signal.

(2) The downlink synchronization signal is used for the time and frequency synchronization UE and eUTRAN during cell search. The necessary prerequisite for UE and eUTRAN to do business connection is the synchronization of time slot and frequency. The downlink synchronization signal consists of two parts:

① Main synchronous signal.

② Auxiliary synchronous signal.

3. Uplink physical channel

The uplink transmission mechanism is similar to that of the downlink. When UE needs to transmit information to eNodeB, information is transmitted through physical channel and reference signal. LTE contains three uplink physical channels, which are:

(1) Physical Uplink Shared Channel, PUSCH.

(2) Physical Uplink Control Channel, PUCCH.

(3) Physical Random Access Channel, PRACH.

4. Uplink physical signal

The uplink physical signal includes:

(1) Dedicated Reference Signal, DRS.

(2) Sounding Reference Signal, SRS.

Task 5   HARQ Technology

HARQ is the short form of Hybrid Automatic Repeat request.Channel scheduling and rate control can be achieved by using the fast fading characteristics of wireless channel, but there are always some unpredictable interference that leads to signal transmission failure. Therefore, FEC (Forward Error Correction) technology is needed. The basic principle of FEC is to increase redundancy in the transmission signal, that is, to add parity bits in the information bits before the signal transmission. The check bits are obtained by calculating the information bits using the method determined by the coding structure. In this way, the number of bits transmitted in the channel will be greater than the number of original information bits, thus introducing redundancy in the transmission signal.

Another way to solve the transmission error is to use ARQ (Automatic Repeat Request) technology. In ARQ scheme, the receiver judges the correctness of the received packet by error detection (usually CRC check). If the packet is judged to be correct, then the received data is error free and the transmitter is informed by the sent ACK (Acknowledge Character) response information. If the packet is judged to be wrong, the transmitter will be informed by the sent NACK (Negative ACKnowledge) response information, and the transmitter will send the same information again.

Most communication systems combine FEC with ARQ, which is called Hybrid ARQ or HARQ. HARQ uses FEC to correct part of all errors, and judges uncorrectable errors through error detection. The packet received in error is dropped and the receiver requests to resend the same packet.

HARQ can be divided into adaptive and non-adaptive. Adaptive HARQ is that some or all attributes of the initial transmission can be changed during retransmission, such as modulation mode, resource allocation, etc., and the change in these attributes requires additional signaling notification. Non adaptive HARQ means that the attributes changed during retransmission are negotiated between the transmitter and the receiver, and no additional signaling notification is required.

The downlink of LTE adopts adaptive HARQ, while the uplink supports both adaptive and non-adaptive HARQ. The non-adaptive HARQ is only triggered by the NACK response information carried in the PHICH channel; the adaptive HARQ is realized through PDCCH scheduling, that is, the base station does not feed back NACK after finding the received output error, but schedules the parameters used for retransmission through the scheduler.

Task 6   Cell Search

1. Time-frequency resource

LTE system uses primary and secondary synchronous signals for cell search. For frame structure 1, the primary/secondary synchronization signal is mapped to 72 subcarriers in the middle of the last/penultimate OFDM symbol of time slot 0 and time slot 10, as shown in Fig. 2-6-1 (taking general CP as an example).

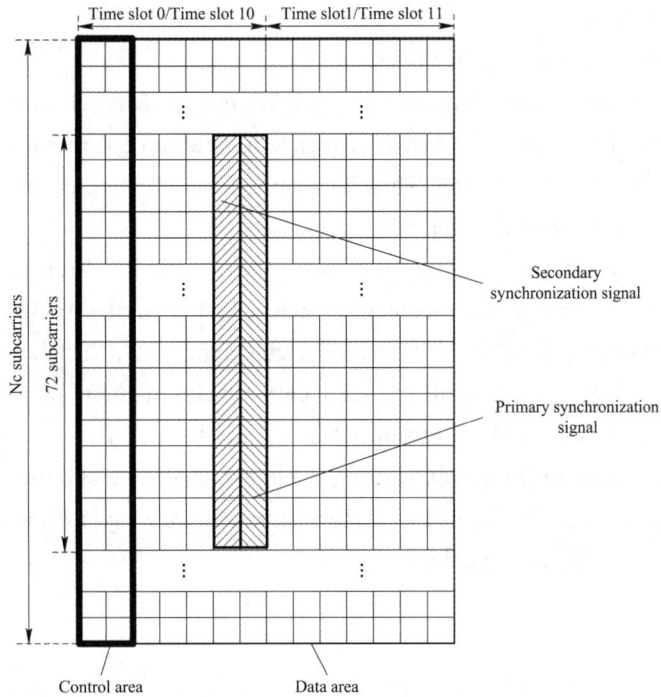

Fig. 2-6-1   Primary / secondary synchronous signal position(frame structure 1, general CP)

For frame structure 2, the primary synchronization signal is mapped to the middle 72 subcarriers on the first OFDM symbol in DwPTS; the secondary synchronization signal is mapped to the middle 72 subcarriers on the last OFDM symbol in time slot 1 and time slot 11, as shown in Fig. 2-6-2 (taking general CP as an example).

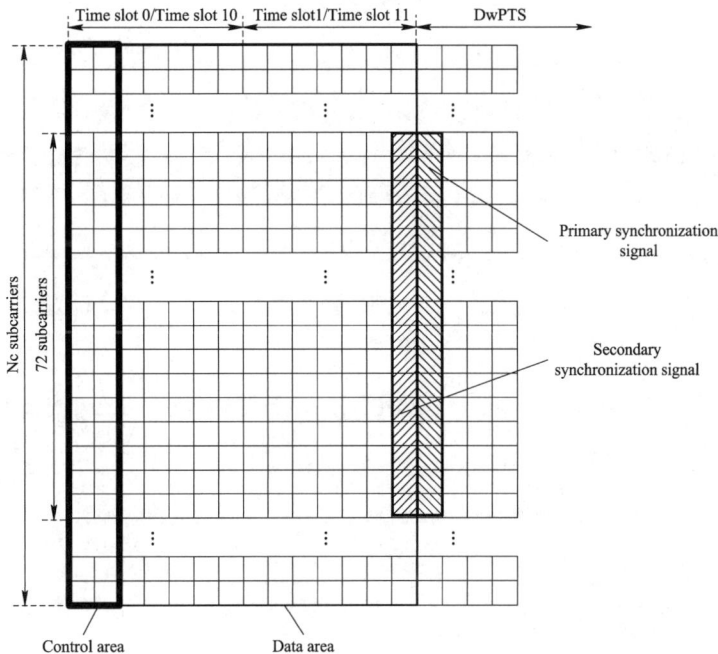

Fig. 2-6-2   Primary / secondary synchronous signal position (frame structure 2, general CP)

2) Cell search process

The cell search process of LTE includes:

(1) Obtain 5ms timing and physical layer cell ID from P-SCH (Primary Synchronization Channel).UE uses the basic synchronization sequence to determine symbol synchronization and identify the cell ID in a cell group. This is obtained by UE comparing three local basic synchronization sequences with the received signals.

(2) Obtain the wireless frame timing and cell group ID from S-SCH (Secondary Synchronous Channel). UE uses the secondary synchronization signal to determine the wireless frame timing and cell group ID index of the cell detected symbol synchronization in the first step, which is obtained by UE comparing all the secondary synchronization signal with the received signals; according to the detected cell group ID index and the cell ID in the cell ID group, UE obtains the cell ID; while detecting S-SCH, the CP length can also be obtained by detection.

(3) Verification of cell ID obtained from downlink reference signal (optional).

(4) Read BCH (Broadcast Channel).

Training Module 3　Network Test

【Brief Description】

The use of Outum 8.10 and network test method are focused in this module. Outum 8.10 integrates the foreground with the background. The foreground collects and monitors wireless data in real time, when the background makes an statistical analysis. Outum 8.10 guides users to locate the problems in wireless environment, and evaluates and monitors the performance of wireless network in real time.

【Training Elements】

1. Knowledge objective

(1) Knowledge about Outum 8.10.

(2) Network test.

2. Capacity objective

(1) Master the usage of Outum 8.10.

(2) Master the network test method. Use testing tools and wireless technology for network testing and optimization.

【Training Requirements】

1. Preparation of tools, instruments and equipment

Outum 8.10 software, mobile phones for testing and computers.

2. Knowledge appraisal point

(1) The usage of Outum 8.10.

(2) Network test method.

3. Skill assessment point

(1) Be able to connect test equipment and use test software correctly.

(2) Make the analysis of test data.

Task 1　Install and Start Software

Outum 8.10, a network testing software, supports TD-SCDMA/TDD-LTE networks. The location of the product in the whole TDD-LTE network is shown in Fig. 3-1-1.

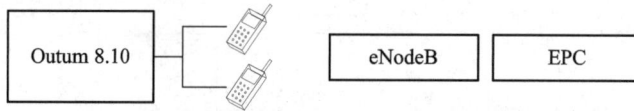

Fig. 3-1-1   Location of Outum 8.10

1. Software installation

The Outum 8.10 is simple to install. Following the prompts of the installation program and the recommended installation options, the software installation can be completed.

2. Software startup

After the Outum 8.10 is installed, it can be started in two ways below:

(1) Start from the start menu: select "Start→program→LTE_FGS→LTE_FGS".

(2) Launch from desktop shortcut: double click the shortcut "LTE_FGS" on the desktop.

After starting the software, wait a moment for the software to open, and the user interface of the software operation is then presented.

Task 2   User Interface

The Outum 8.10 user interface includes navigation bar, work area view, menu bar, tool bar and title bar, status display bar, etc. Outum 8.10 user interface is shown in Fig. 3-2-1.

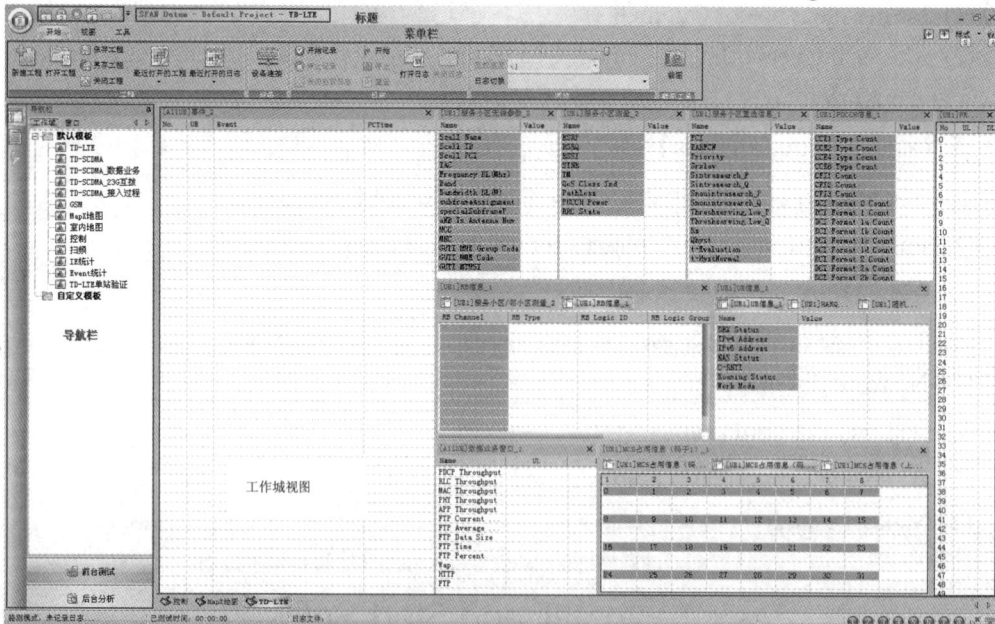

Fig. 3-2-1   Outum 8.10 user interface

1. Navigation bar

The navigation bar includes work area bar, IE bar, event bar and resource manager. The work area and window tree in the work area bar are respectively shown in Fig. 3-2-2 and Fig. 3-2-3.

Fig. 3-2-2   Work area

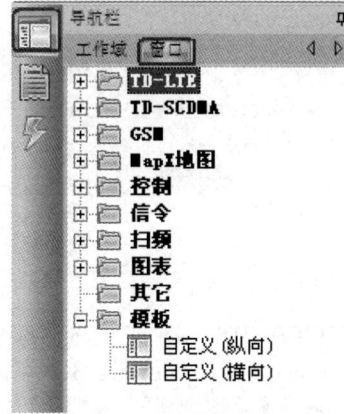

Fig. 3-2-3   Window tree

The IE bar includes standard IE and Common IE, as shown in Fig. 3-2-4.

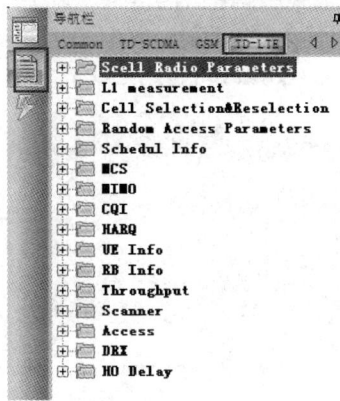

Fig. 3-2-4   IE bar

The event bar includes standard events(TD-SCDMA event，GSM event，TD-LTE event), Common event, EvevtDef (custom event), Application event and Script event. The event bar is shown in Fig. 3-2-5.

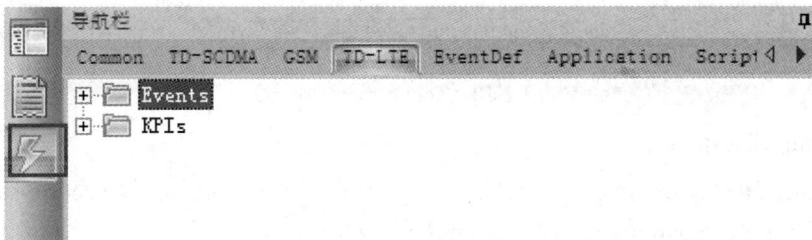

Fig. 3-2-5   Event bar

The resource manager can switch between foreground test and background analysis, as shown in Fig. 3-2-6.

Fig. 3-2-6   Resource manager

2. Work area view

The work area view page can display any number of work areas, and can switch to display work areas. The work area view is shown in Fig. 3-2-7.

Fig. 3-2-7   Work area view

3. Menu bar

The foreground road test and background analysis in the tool bar are separated. The foreground menu bar is shown in Fig. 3-2-8.

Fig. 3-2-8   Foreground menu bar

The background analysis bar is shown in Fig. 3-2-9.

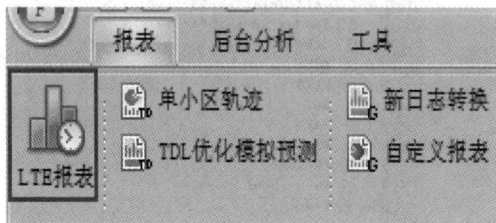

Fig. 3-2-9   Background analysis bar

4. Tool bar and title bar

The tool bar mainly displays shortcuts of some functions, convenient for users to operate, and customizes related operations; the title bar displays the name of current project and the work field of current page. They are shown in Fig. 3-2-10.

Fig. 3-2-10   Tool bar and title bar

5. Status display bar

The status display bar displays the working status and equipment connection status of the current software, as shown in Fig. 3-2-11 and Fig. 3-2-12.

Fig. 3-2-11   Working status bar

Fig. 3-2-12   Equipment connection status bar

Task 3   Project Management

For the project management of Outum 8.10, the specific operation steps are as follows.

1. Create a new project

After starting Outum 8.10, the default state is the last project saved by the user. If "default project.fgs" is penned for the first run, click "new project" on the "tool bar" to create a new project. The dialog box of new project(新建工程) is shown in Fig. 3-3-1.

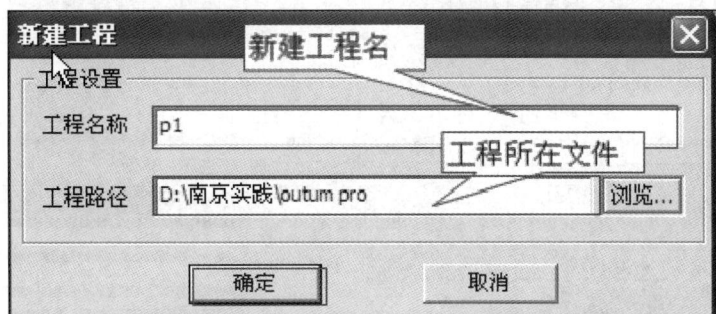

Fig. 3-3-1   Dialog box of new project

Click "OK(确定)" to create project p1. The new project is shown in Fig. 3-3-2.

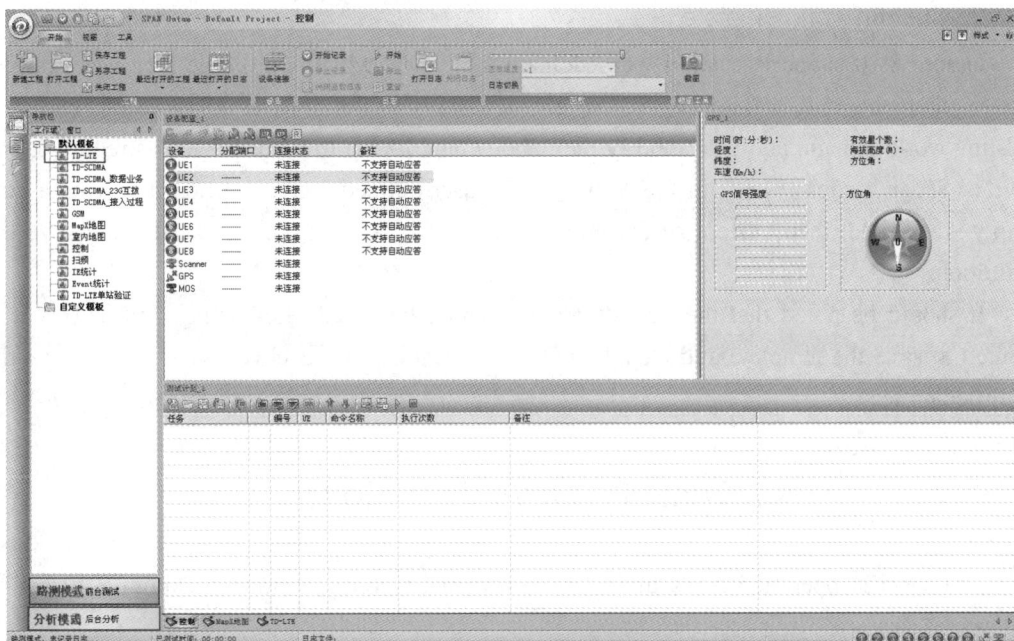

Fig. 3-3-2   Create a new project

Open TD-LTE work area in "navigation bar→work area→default template", as shown in Fig. 3-3-3.

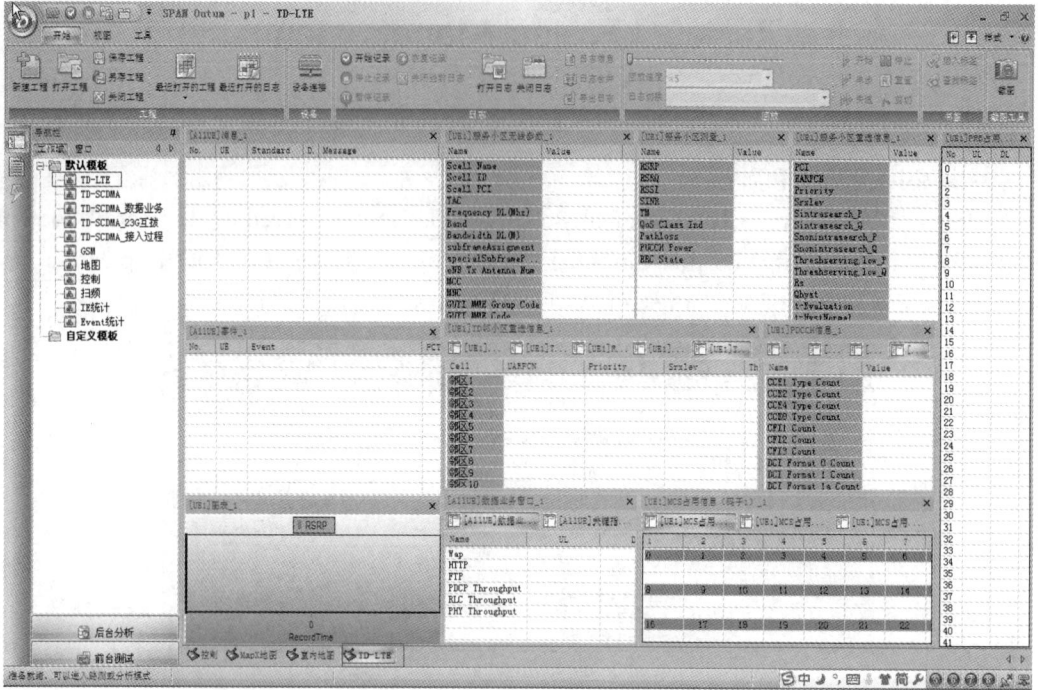

Fig. 3-3-3   TD-LTE work area

2. Custom work template

Outum 8.10 supports custom work domain templates, which can be defined according to test requirements and preferences. There are two ways to define a template. First, copy the existing work domain, and then modify the window and layout to be displayed as required, and save it. Second, click "New" in the "work field - user defined" template item to create a new empty work field, then customize the work field on the basis of the test needs, and save the definition.

It should be noted that the default project cannot be saved after being modified. If it is wanted to save the template and change under the default project, click "save as'" to save the modified project.

Task 4　Drive Test Mode

The specific operation steps of the drive test mode configuration of Outum 8.10 are as follows.

1. Equipment configuration

Click "equipment configuration" in the "control→equipment configuration" window or click "equipment connection(设备连接)" in the tool bar to configure and manage the terminal equipment. The equipment configuration is shown in Fig. 3-4-1.

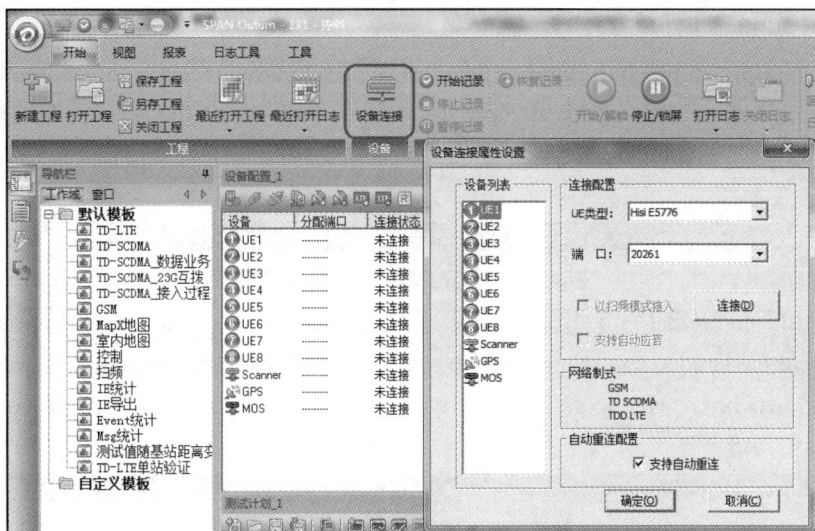

Fig. 3-4-1　Equipment configuration

Select the test terminal model to be connected from the "UE type" drop-down box. At present, LTE only supports Hisilicon and United Microchip terminals. The main models are Qualcomm, Hisi Balong 710, Hisi E398, and LC5160, and the connection of up to 8 test terminals is supported. The connection of Hisilicon LTE terminal is shown in Fig. 3-4-2.

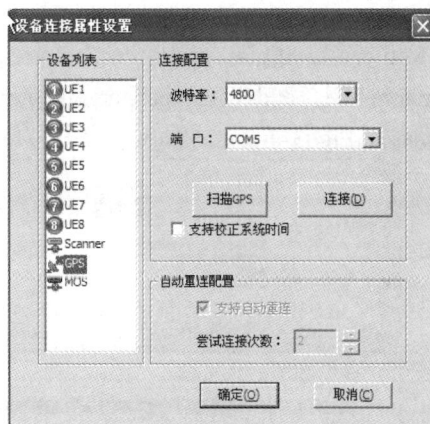

Click "Connect(连接)" to connect the device. After the connection, the label will turn green. If it is needed to disconnect, click "Disconnect(断开)". After connecting the test terminal, connect the GPS. Select the baud rate of 4800, then click "Scan GPS (扫描 GPS)" to find the port where the GPS is, and click "Connect(连接)" to connect the GPS. The connection of GPS is shown in Fig. 3-4-3.

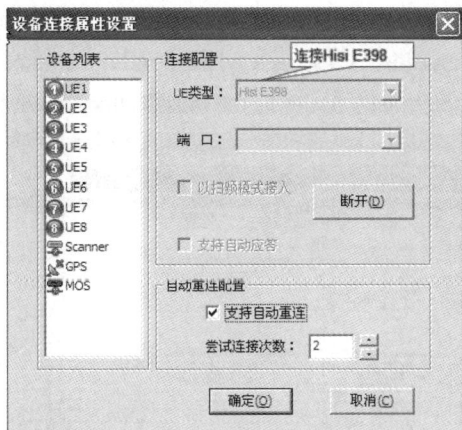

Fig. 3-4-2　Connection of Hisilicon LTE terminal　　Fig. 3-4-3　Connection of GPS

After the terminal is connected, the interface begins to receive the data reported by the terminal.

Other terminals can be connected with the same steps. At present, MOS and scanner functions are not realized.

2. Test script configuration

The test script can be configured in the "test plan" window of the "control" work area. The following functions are realized: create new script, open script, save log script, save log script as, add command, delete command, edit command, and start running and stop running. Test script is shown in Fig. 3-4-4.

Fig. 3-4-4   Test script configuration

Click "add command" on the tool bar of "test plan" window to open command editing window, where the "added command" can be edited. For "Power On" command, select "UE1" in "UE" drop-down box, "Hisilicon" in "UE type", and then click "add" to add command, as shown in Fig. 3-4-5.

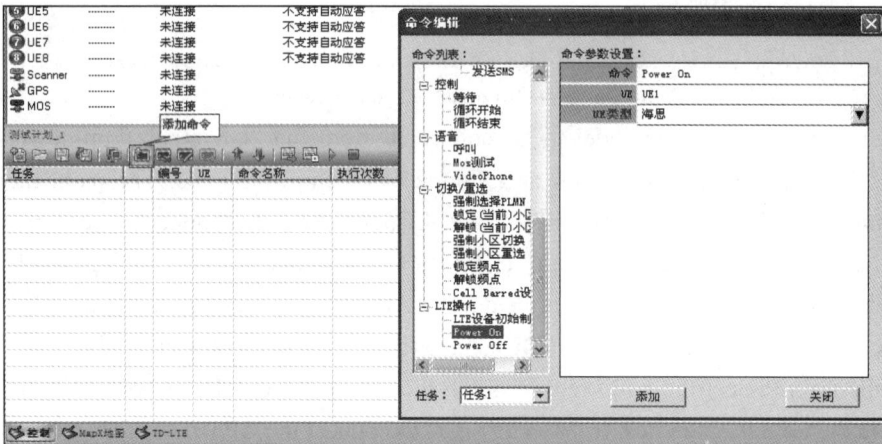

Fig. 3-4-5   Add command

If you need to connect multiple terminals and execute commands in parallel at the same time, you can configure multi-terminal script commands. Select the corresponding connecting terminal in "UE" and "UE type", and then modify the "task" name in the "task" drop-down box, as shown in Fig. 3-4-6.

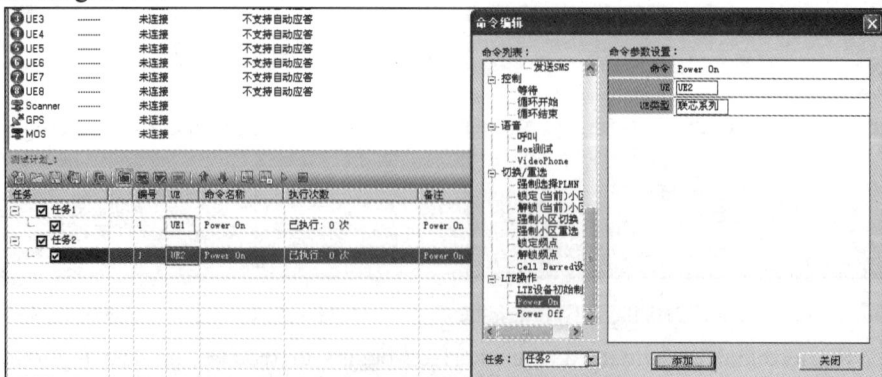

Fig. 3-4-6   Execute commands in parallel

Here are the LTE script commands currently supported by Outum 8.10.

(1) Lock/unlock cell. Set Band and frequency point to lock/unlock cell, as shown in Fig. 3-4-7.

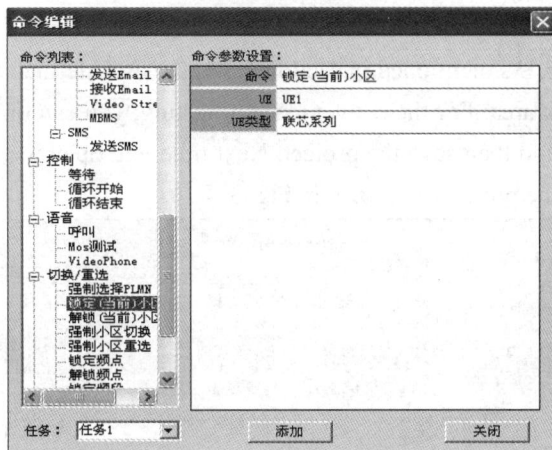

Fig. 3-4-7   Lock/ unlock cell

(2) Lock/unlock frequency point. Set Band and frequency point to lock/unlock frequency point, as shown in Fig. 3-4-8.

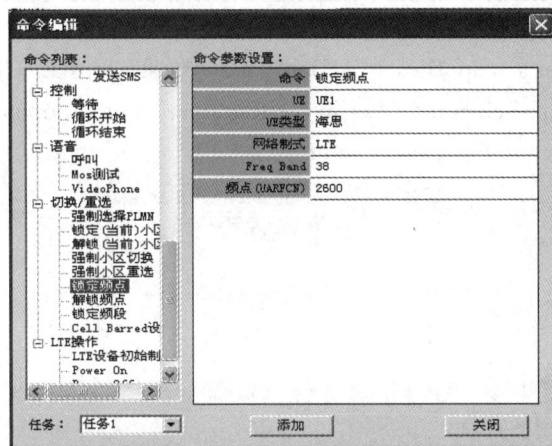

Fig. 3-4-8   Lock/unlock frequency point

(3) Lock the frequency band. Set band to lock the frequency band, as shown in Fig. 3-4-9.

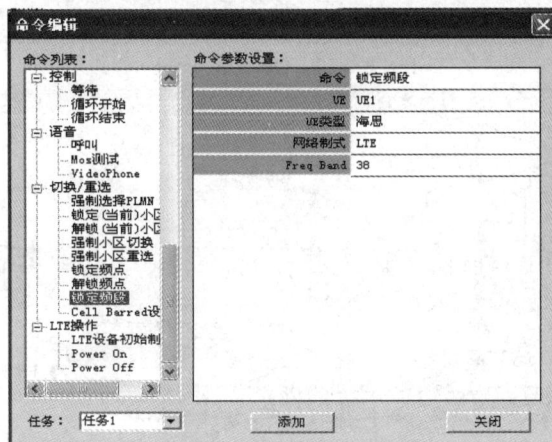

Fig. 3-4-9   Lock the frequency band

3. Map parameters

(1) Import the map. As mentioned in section 2.3.1, the user-defined work template can also be used in the map work area. For the convenience of testing, you can drag the relevant window to the same work area, and then save the project. Next time you open the project, you don't need to reconfigure it. Open the protect, as shown in Fig. 3-4-10.

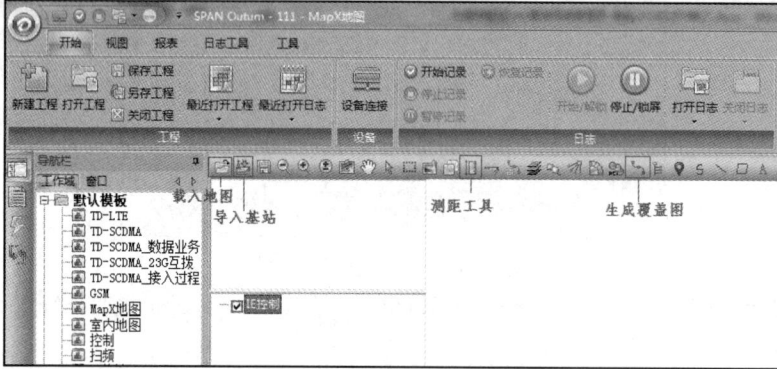

Fig. 3-4-10　Open the project

Outum 8.10 supports map files in "*.tab" and "*.gst" format. Click "open", the layer selection window pop-up. Select the layer in the test area, and then open it. Open the layer and import the layer, as shown in Fig. 3-4-11 and Fig. 3-4-12.

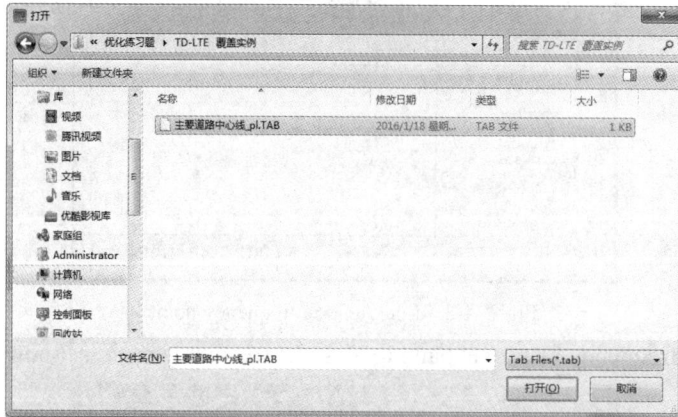

Fig. 3-4-11　Open the layer

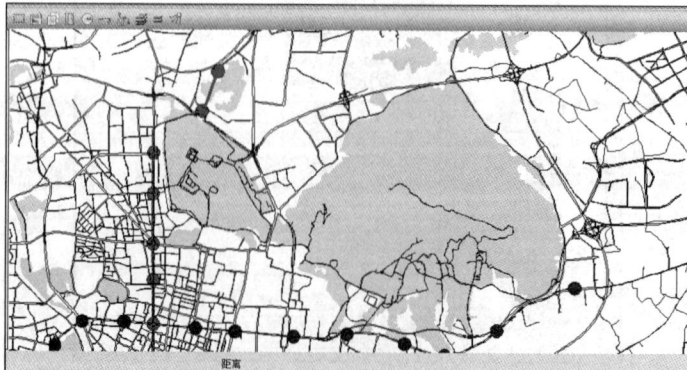

Fig. 3-4-12　Import the layer

(2) Import the base station data table. Outum 8.10 only supports the import of base station table in *.xls/*.xlsx format, and the names of key parameters should be in English. The base station data table is shown in Table 3-4-1.

Table 3-4-1　Base station data table

Name	Meaning	Required field	Note
Azimuth	Direction angle	Y	
BeamWidth	Beam width	Y	
CellName	Cell name	Y	
Frequency DL	Center frequency point	Y	This field is the center frequency point, such as 2600
ECI	Cell ID	Y	Unique identification of a cell under PLMN
eNodeB-ID	Base station number	Y	Unique identification of a base station under PLMN
High	Station height	Y	
Latitude	Latitude	Y	
Longitude	Longitude	Y	
MCC	Mobile country code	Y	
MME group ID	MME group number	N	Whether cross MME switching is reflected
MMEC	MME number	N	Whether cross MME switching is reflected
MNC	Mobile network number	Y	
PCI	Physical cell ID	Y	
RsPower(dBm)	CRS reference power	N	
SiteName	Base station name	Y	
TAC	Tracking area	Y	Unique identification of a TA under PLMN
tilt_total	Total declination	Y	
本地小区标识(Local cell identification)	Local cell ID	N	
带宽(Bandwidth)	System bandwidth	N	
室内外(Indoor and outdoor)	Base station type	N	Distinguish between macro station and indoor station
站型(Station type)	Base station type	N	S111, O1, etc.

For details, please refer to the base station table template below. You can add or delete relevant fields according to the test requirements, as shown in Table 3-4-2.

Table 3-4-2   Base station table template

MCC	MNC	eNodeB-ID	SiteName	CellName	ECI	TAC	Frequency_DL	PCI	Longitude	Latitude	RsPower(dBm)	High	Azimuth	BeamWidth	tilt_total
460	00	144	江宁移动 (Jiangning Mobile)	江宁移动 (Jiangning Mobile)-1	36864	511	2600	60	118.841440	31.937750	15	58	0	60	8
460	00	144	江宁移动 (Jiangning Mobile)	江宁移动 (Jiangning Mobile)-2	36865	511	2600	61	118.841440	31.937750	15	58	120	60	8
460	00	144	江宁移动 (Jiangning Mobile)	江宁移动 (Jiangning Mobile)-3	36866	511	2600	62	118.841440	31.937750	15	58	240	60	8

Click "import BTS table" to open the "BTS table import" dialog box, as shown in Fig. 3-4-13. Select "TD-LTE" in the "system", select the file path of the BTS table, and click "OK".

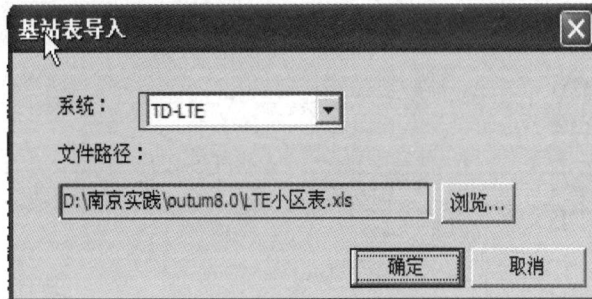

Fig. 3-4-13   Import BTS table

Next step is to map the fields, as shown in Fig. 3-4-14.

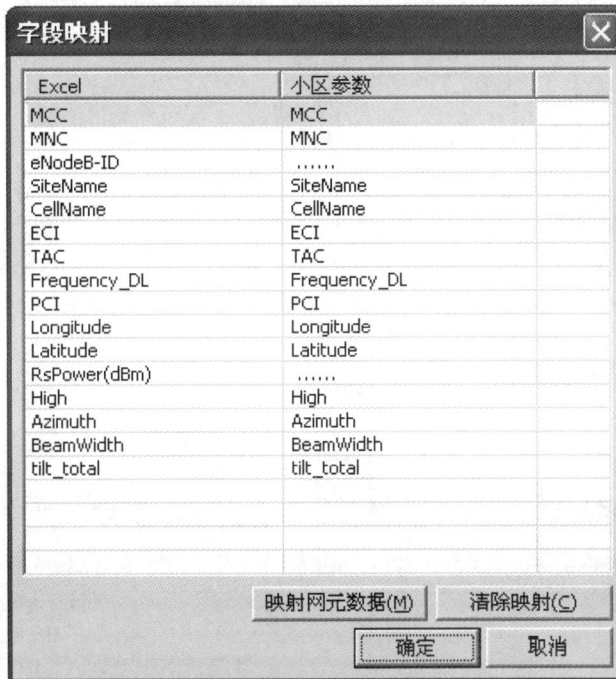

Fig. 3-4-14   Map the fields

After confirmation, it will be prompted that the BTS file is loaded successfully and that the cell layer is generated. Click "OK" and the import is completed, as shown in Fig. 3-4-15.

Fig. 3-4-15   Load base station file

The cell layer generated from the above steps is shown in Fig. 3-4-16.

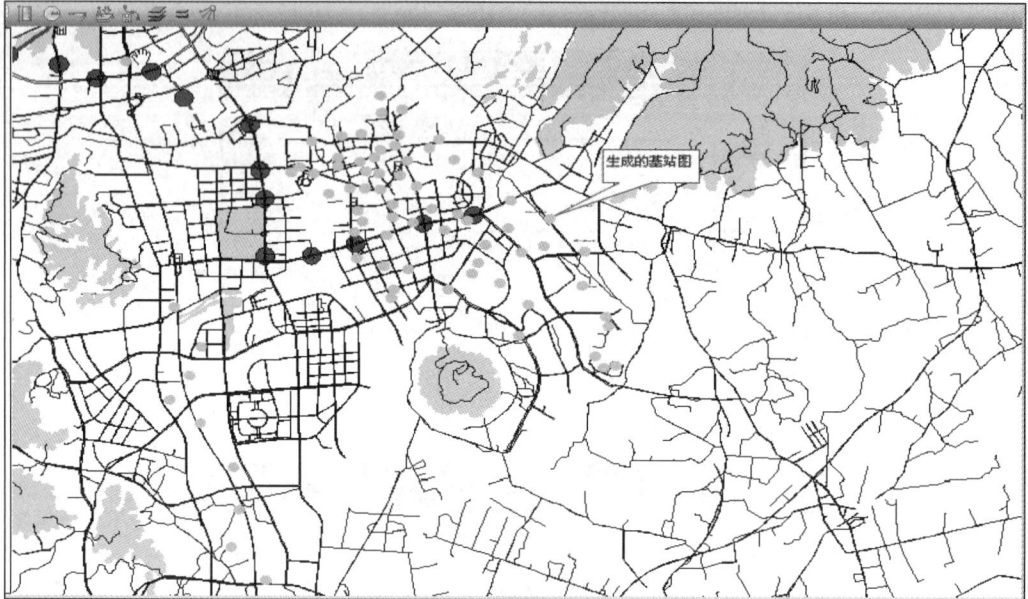

Fig. 3-4-16   Cell layer

Right click "layer setting" in the map window to open the layer control window, where you can edit the layer. Click "Add(增加)" to add the layer, and "Delete(删除)" to remove the selected layer. To display the attributes of a layer, select "mark", and then configure it in the mark option window, as shown in Fig. 3-4-17. After the configuration is completed, click "OK", and then check "auto mark(自动标注)" to confirm.

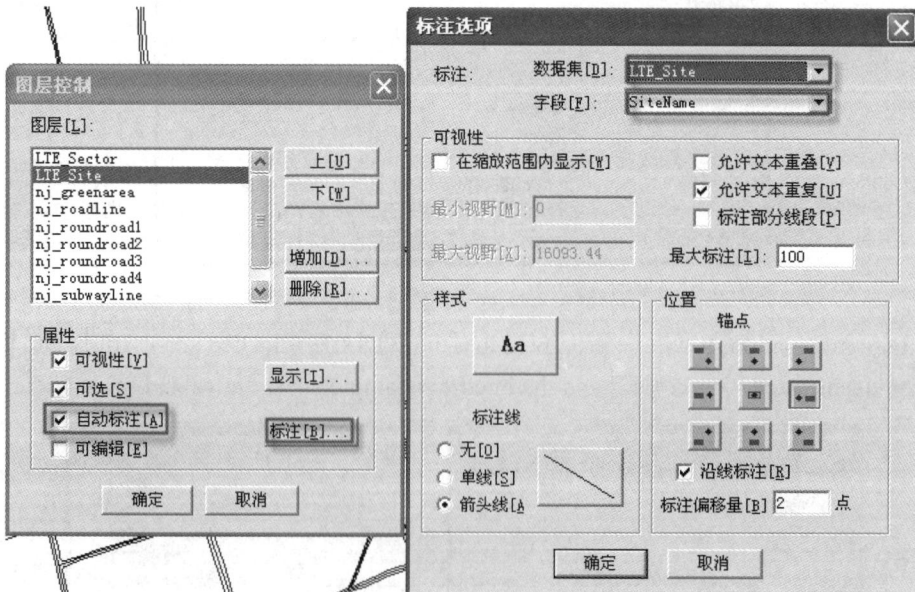

Fig. 3-4-17   Layer setting

The marked layer result is shown in Fig. 3-4-18.

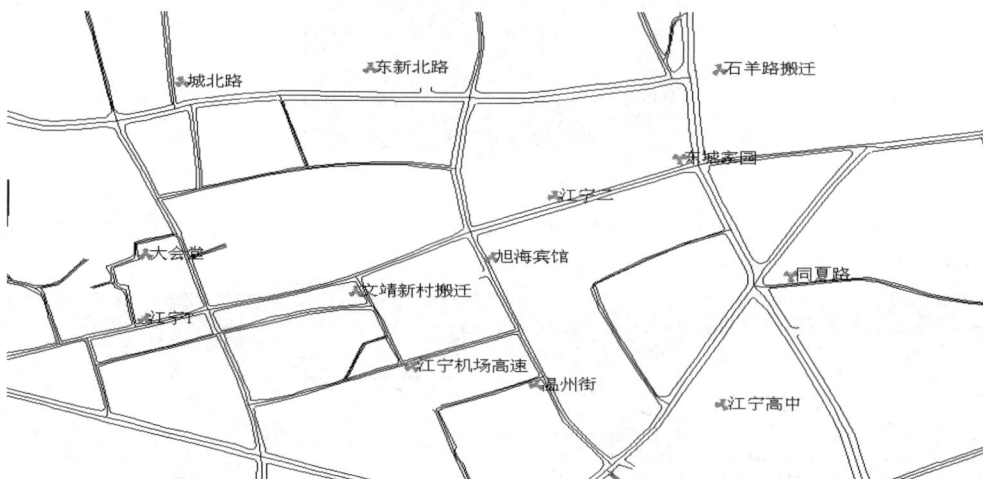

Fig. 3-4-18　Marked layer result

4. Start recording the log

(1) Outdoor drive test. In the tool bar or shortcut menu bar, click "start recording(开始记录)", then select outdoor test as the test type, click "OK", then select the saving path and log name, and click "save(保存)" to start recording the log, as shown in Fig. 3-4-19.

Fig. 3-4-19　Start recording

To stop the test, click "stop recording(停止记录)", as shown in Fig. 3-4-20.

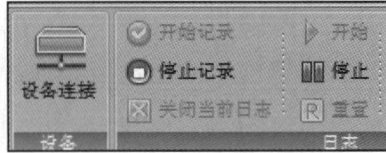

Fig. 3-4-20　Stop recording

After recording the log is stopped, you can view the terminal reporting parameters in the window of the work domain, as shown in Fig. 3-4-21.

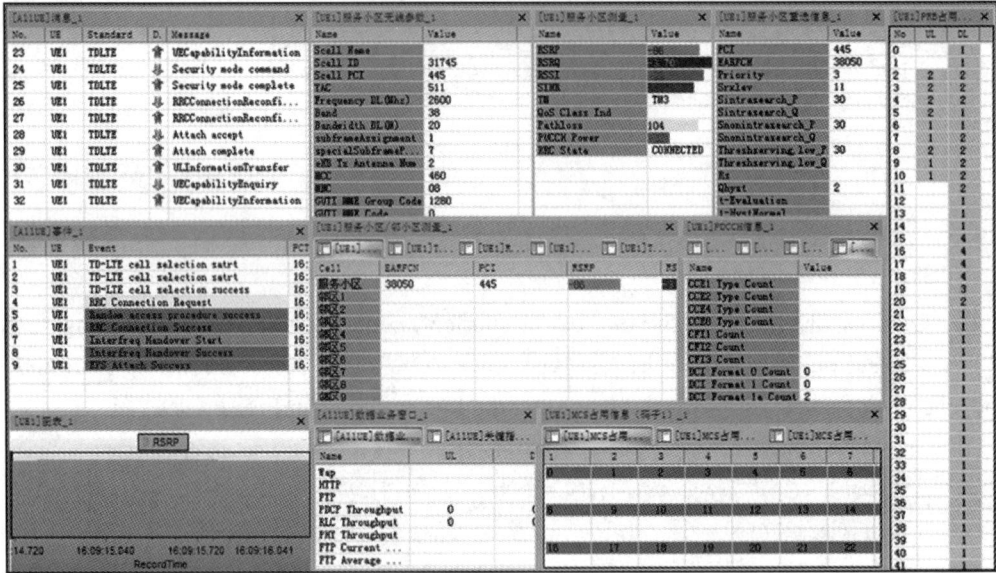

Fig. 3-4-21　View the terminal reporting parameters

Click the connection tool " " to view the pull wire of the whole network, and click it again to remove the pull wire of the whole network, as shown in Fig. 3-4-22.

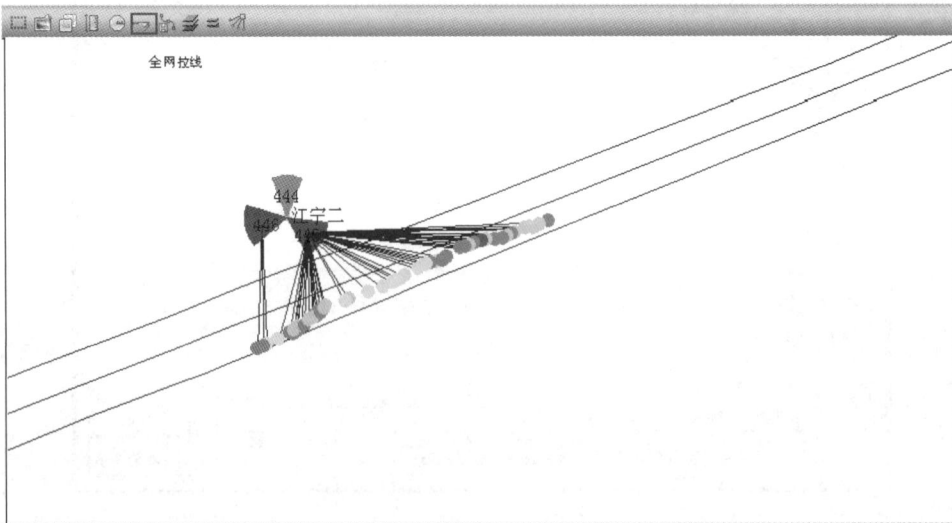

Fig. 3-4-22　View the pull wire of the whole network

Click "", and then click the base station sector in the map to view the pull line of the sector, as shown in Fig. 3-4-23.

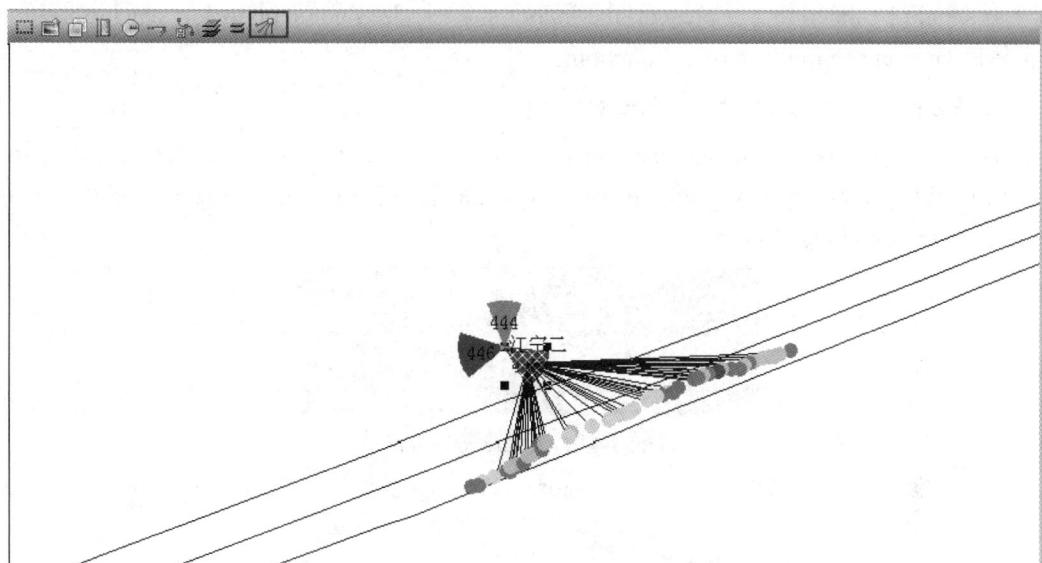

Fig. 3-4-23   View the pull line of the sector

(2) Indoor test. The steps are as follows.

① Open the indoor map work area in the navigation bar, and import the map layer, as shown in Fig. 3-4-24.

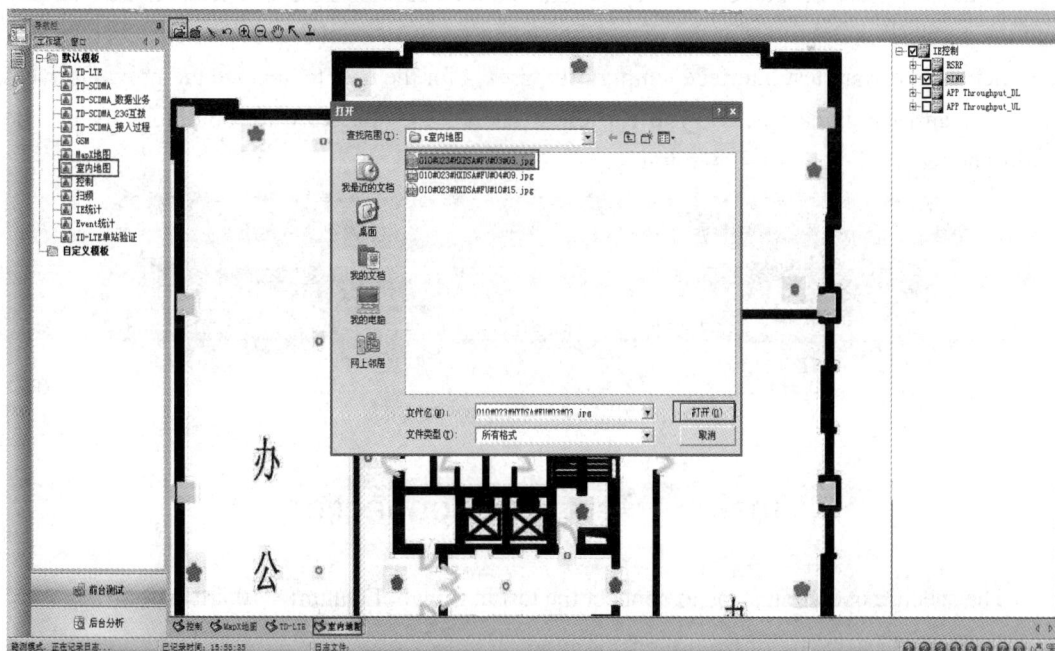

Fig. 3-4-24   Open the indoor map work area

② Connect the test terminal (e.g. mobile phone, GPS) and start recording the log. (It should be noted to select outdoor drive test when selecting storage log type).

③ Mark the track in the map, and support the RSRP (Reference Signal Receiving Power) and SINR (Signal to Interference plus Noise Ratio) map to list and mark the points.

It should be noted that at present, the indoor site test does not support log synchronization analysis, layer annotation and other functions.

5. Pause and resume recording the log

After the test is started and the log has been recorded, if it is necessary to pause the recording thlog in emergency, click "pause recording(暂停记录)" after the recording the log is started, as shown in Fig. 3-4-25.

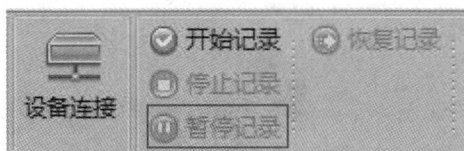

Fig. 3-4-25　Pause recording

To continue to record the log, click "restore recording(恢复记录)", as shown in Fig. 3-4-26.

Fig. 3-4-26　Restore recording

In the test process, if an emergency occurs and the cause of the event needs to be analyzed and the wireless environment at the time of the event needs to be checked, you can click "stop", and then the software test interface temporarily freezes for the user to analyze the previous event. After the analysis is completed, you can click "start", and the software continues to test the status and record the log, as shown in Fig. 3-4-27.

Fig. 3-4-27　Stop testing

Task 5　Terminal Connection

The specific operation steps to connect the test terminal of Outum 8.10 are as follows.

1. United Microchip terminal connection

(1) After the installation of Outum 8.10, a "LTE_FGS" shortcut appears in the desktop and start menu. Click the shortcut to open the "LTE_FGS" software and enter the control domain interface, as shown in Fig. 3-5-1.

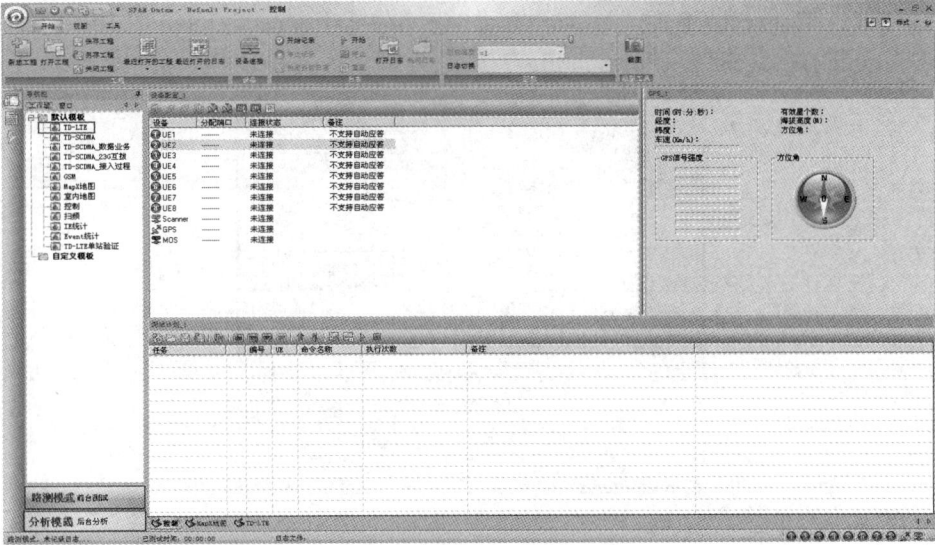

Fig. 3-5-1　Control domain interface

(2) Click "device connection", select the corresponding UE type according to the actual terminal model, and select the COM port connected to the terminal, as shown in Fig. 3-5-2.

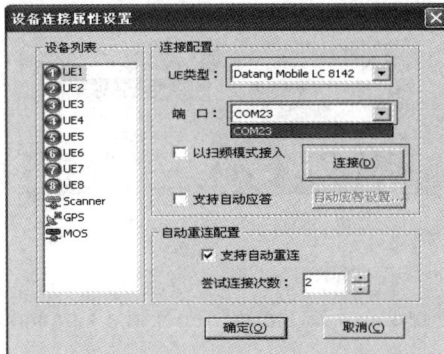

Fig. 3-5-2　Device connection

2. HUAWEI E5776s connection

(1) Select "Hisi Balong 710" as UE type, as shown in Fig. 3-5-3.

Fig. 3-5-3　Select the UE type

(2) Click "Connect (连接)" to start Hisi UE Agent. Connect Outum 8.10 to HUAWEI E5776s through UE Agent, as shown in Fig. 3-5-4.

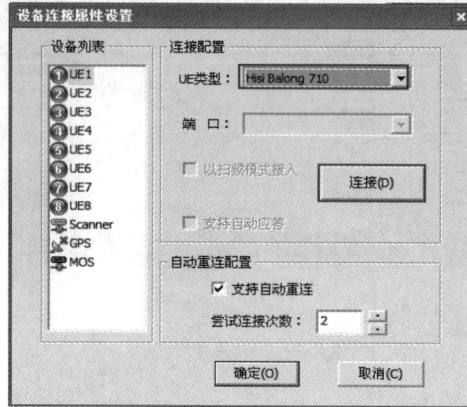

Fig. 3-5-4   connecting through UE Agent

(3) Right click "UE Agent", find "Command", and you can execute the script command, as shown in Fig. 3-5-5.

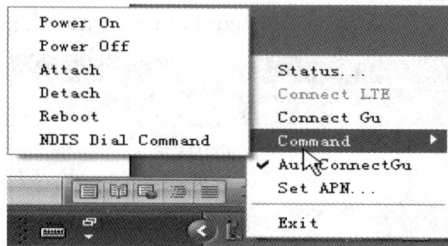

Fig. 3-5-5   Script command

3. Qualcomm terminal connection (L1)

After connecting the L1 terminal to the PC, open the automatic dialing switching tool, as shown in Fig. 3-5-6.

Fig. 3-5-6   Automatic dialing switching tool

Confirm that the equipment is ready, and then click "open (打开)". Under normal circumstances, the following status appears, and L1 terminal restarts. If the terminal cannot restart automatically, please start it manually, as shown in Fig. 3-5-7.

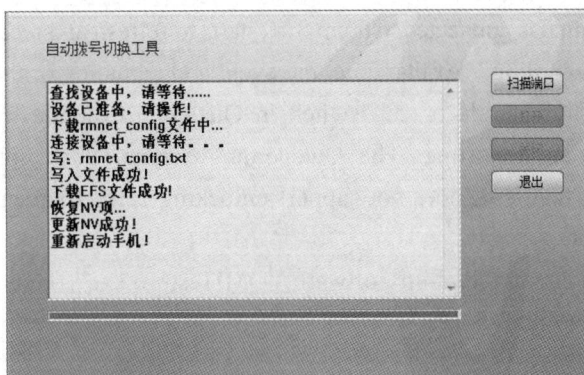

Fig. 3-5-7　Restart automatically

Open the Outum 8.10 and click "device connection".

(1) Check the terminal usage port (Diagnostics 9026) in device manager port, as shown in Fig. 3-5-8.

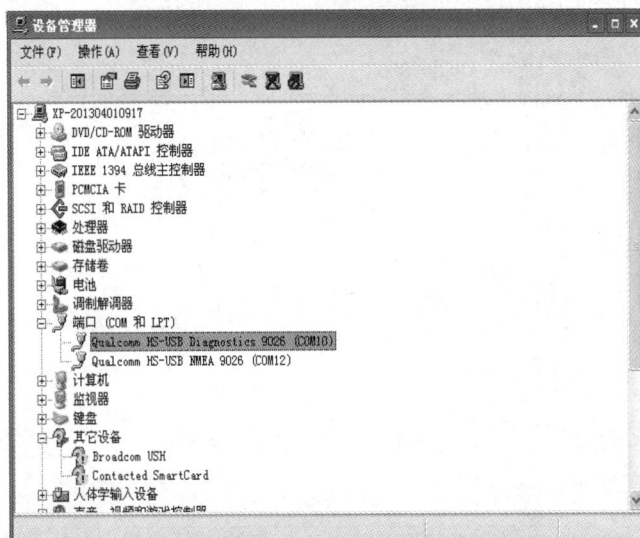

Fig. 3-5-8　Check the terminal usage port (Diagnostics 9026)

(2) To connect the terminal in the Outum 8.10 device connection, select "QualComm" for UE type, and select the part "COM10", and then click "Connect", as shown in Fig. 3-5-9.

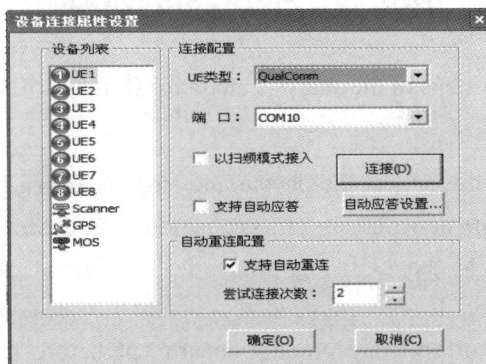

Fig. 3-5-9　Terminal connection

(3) After the terminal is connected, Outum 8.10 starts to refresh and receive the terminal data.

(4) Click "start recording". At this time, the event and signaling in the interface are empty. You need to set the flight mode or cell switch in Qualcomm mobile terminal to trigger the signaling and event data reporting. The Qualcomm terminal does not support the forced command temporarily, that is, it does not support unlocking cell / frequency locking segment / frequency locking point.

(5) Start the Qualcomm dial-up software "QMITestpro.exe", and then make a dial-up connection, as shown in Fig. 3-5-10.

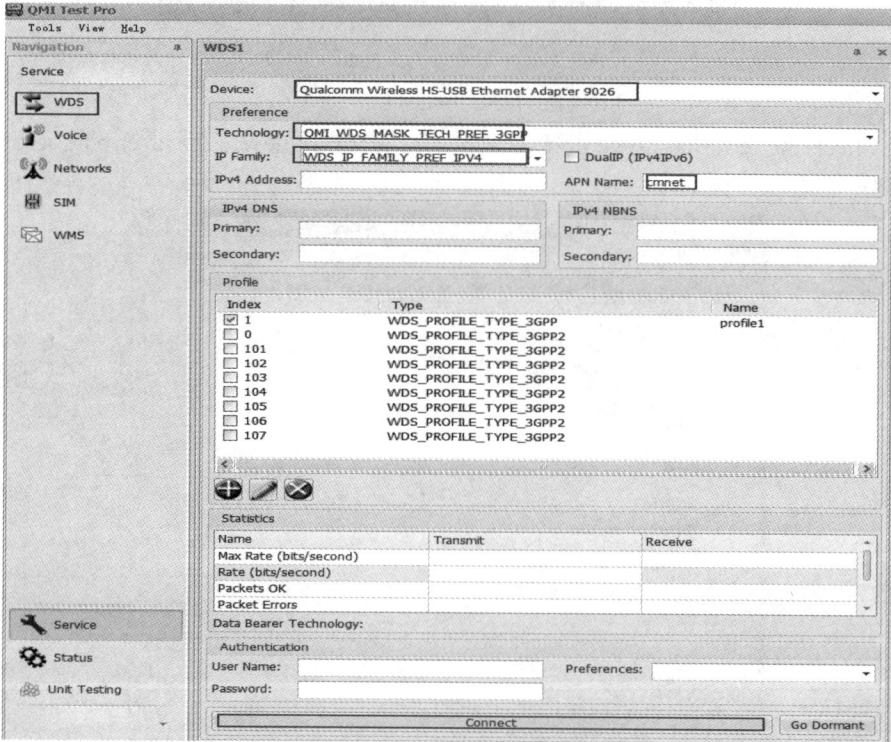

Fig. 3-5-10   Make a dial-up connection

Task 6   Analysis Mode

The specific operation steps of the analysis mode for Outum 8.10 are as follows.

1. Log playback

Before log playback, disconnect the device and stop recording the log. Outum 8.10 only supports log playback of the new version of LGL. During the drive test, the automatic storage mode is the new version of the LGL log, while the log format of the old version can be converted to the new version. It is recommended to obtain the latest version of Outum software from the interface person of software product department before the project starts. The log playback is shown in Fig. 3-6-1.

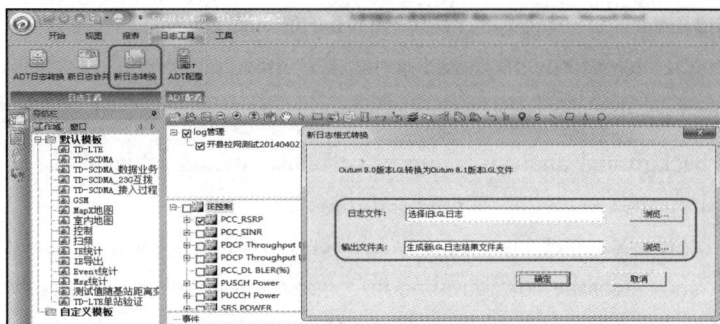

Fig. 3-6-1　Log playback

The LGL type log in the new version supports the analysis and synchronization of single IE. The software supports operations such as log opening, log closing and multi-speed playback. Before opening the log, it is recommended to open the map work area and import the cell base station table to Outum 8.10, so that the map can directly display the whole network cable of cell base station.

Click "Open log", select the log to be played back, and click "Open". After opening the log, select the playback speed and click "Start" to start playback. To stop playback, click "Stop".

The following should be noted in log playback:

(1) Outum 8.10 supports the function of data synchronization for single IE.

(2) Click the IE map track point, and use the left and right keys of the keyboard to carry out the back and forward functions of the log time respectively.

(3) The log analysis module has optimized the playback function. Open the log, and you can map all the log data to Outum 8.10.

(4) Before opening the log, it is recommended to first open the map work area and import the cell base station table to Outum 8.10, so that the map can directly display the whole network cable once the log is opened.

2. Whole network coverage map

Outum 8.10 supports PCI and IE column overlay. The operation steps are as follows.

(1) Import cell base station table or cell base station layer in Tab format into the map, as shown in Fig. 3-6-2.

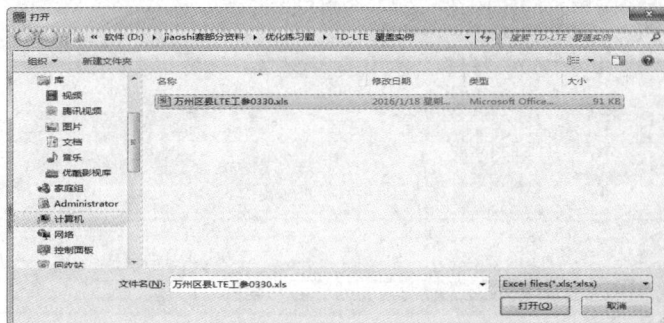

Fig. 3-6-2　Import cell base station table

(2) Open the log in the foreground drive test menu bar.

(3) Select the IE layer to be displayed in the IE control bar.

3. Single cell track coverage

(1) Click "background analysis", import cell base station table in the map or cell base station layer in tab format, and click "single cell track".

(2) Click "browse", select the folder where the LGL log is located, select the location of the output folder, and select the configuration information table of the cell (its format is the same as the cell table format in the statistics report), as shown in Fig. 3-6-3.

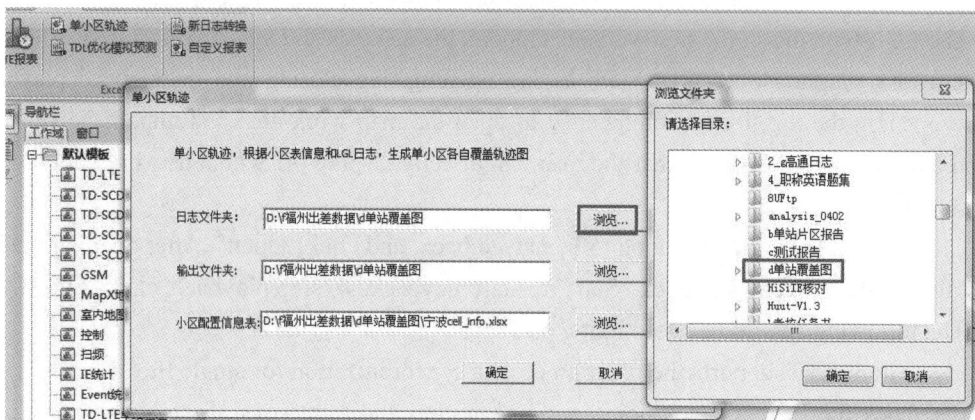

Fig. 3-6-3   Select the configuration information table of the cell

(3) Click "OK" to process the data in the background, as shown in Fig. 3-6-4.

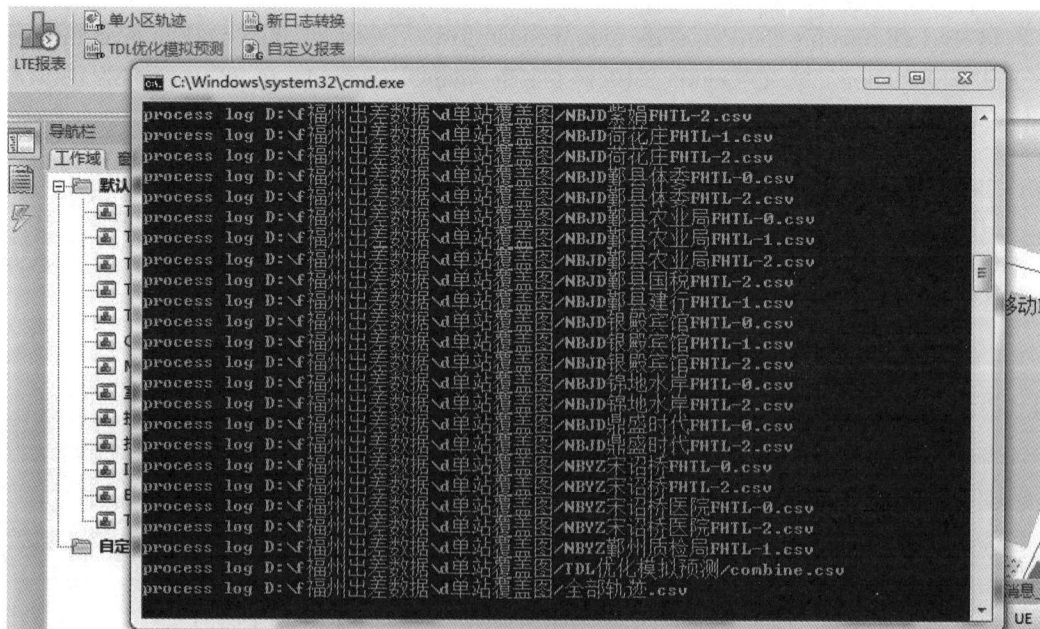

Fig. 3-6-4   Background data processing

(4) The processed data is shown in Fig. 3-6-5.

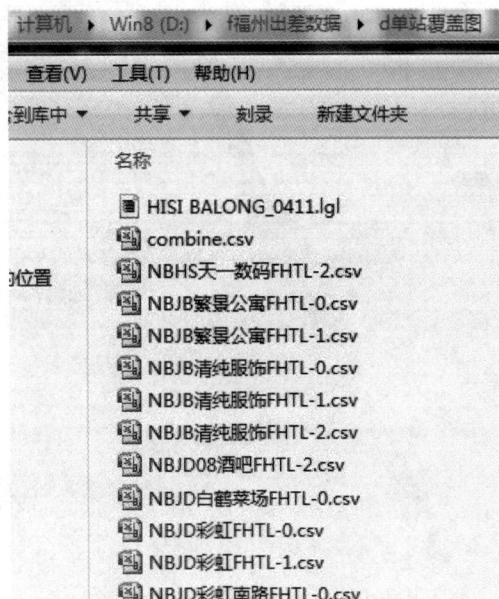

Fig. 3-6-5   Processed data

(5) Click the "overlay" in Fig. 3-6-6, and select the corresponding cell data in the pop-up configuration box.

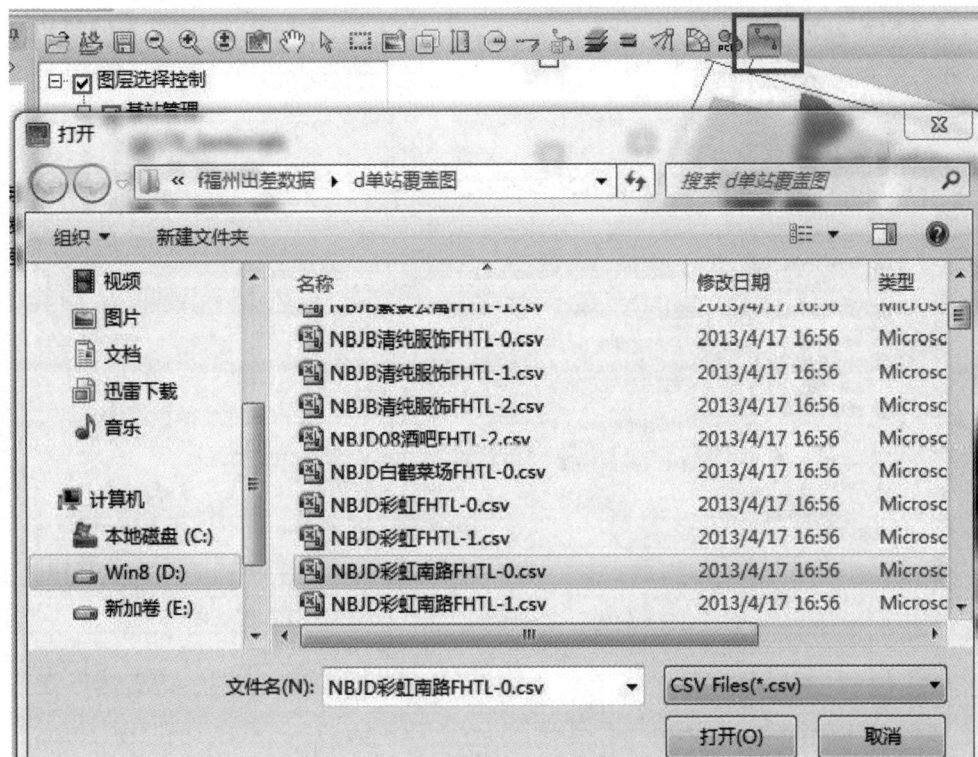

Fig. 3-6-6   Select the corresponding cell data

(6) Click "open", and the single cell track coverage is shown in Fig. 3-6-7.

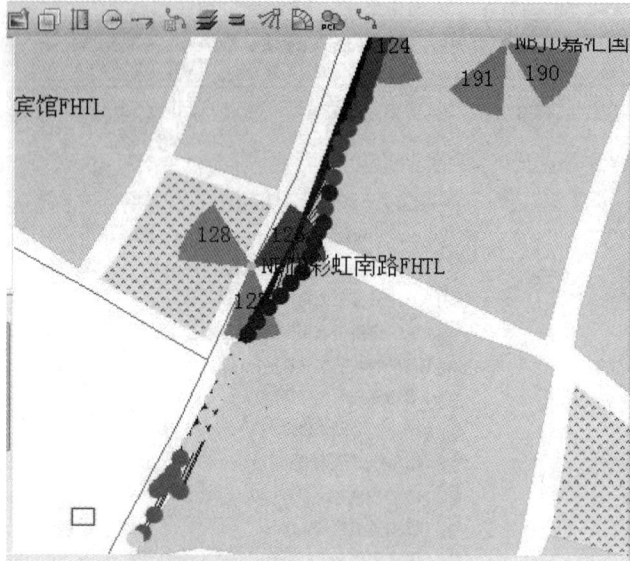

Fig. 3-6-7   Single cell track coverage

4. TDL simulation optimization prediction

(1) Click "background analysis", import cell base station table or cell base station layer in tab format into the map, and click "TDL optimization simulation prediction", as shown in Fig. 3-6-8.

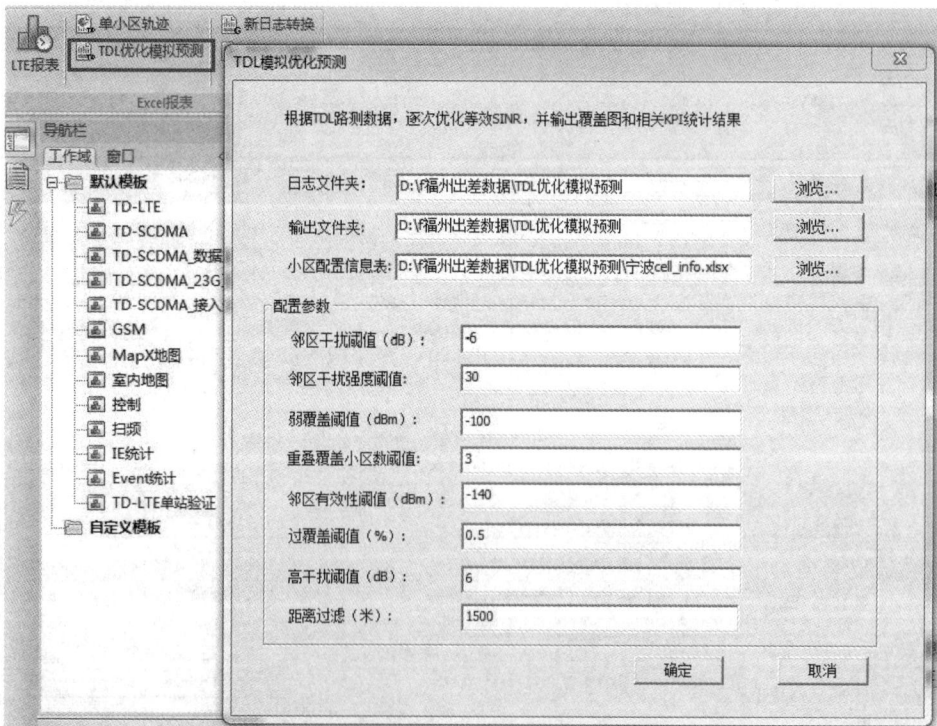

Fig. 3-6-8   TDL optimization simulation prediction

(2) Select LGL log folder in the configuration interface, output path folder, and select cell configuration information table (its format is the same as the cell table format in the statistics report).

(3) Click "OK". The background processing is shown in Fig. 3-6-9.

Fig. 3-6-9   Background processing

(4) Find the generated result file "result" in the output folder, as shown in Fig. 3-6-10.

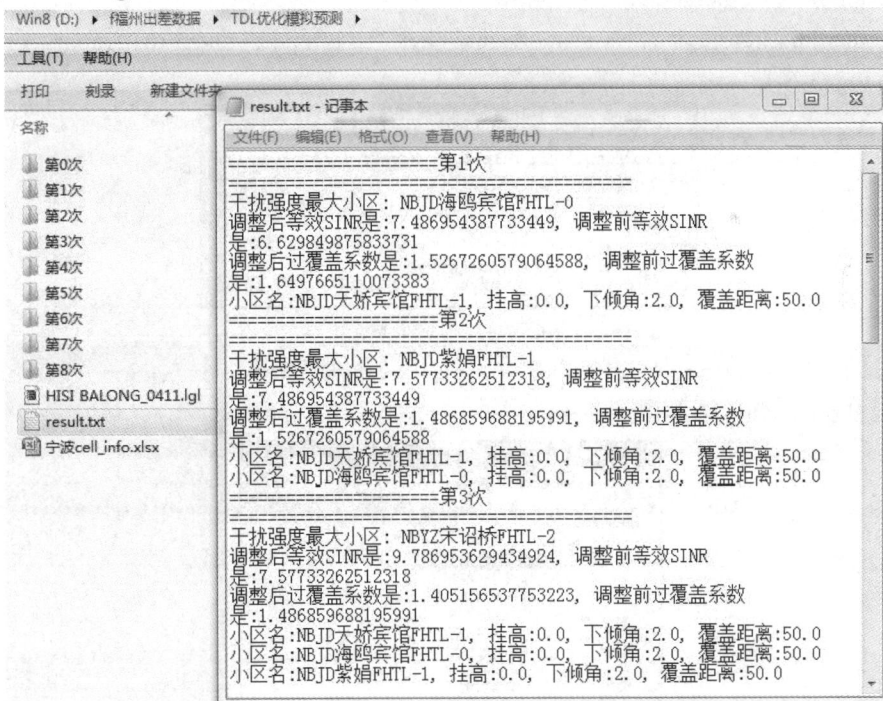

Fig. 3-6-10   File of the generated result

5. Overlapping coverage analysis

At present, Outum 8.10 only supports the output of overlapping coverage table, i.e, it outputs a CSV named "overlap", which has not been mapped yet, and subsequent analysis has to be made through the use of Mapinfo and other tools, as shown in Fig. 3-6-11.

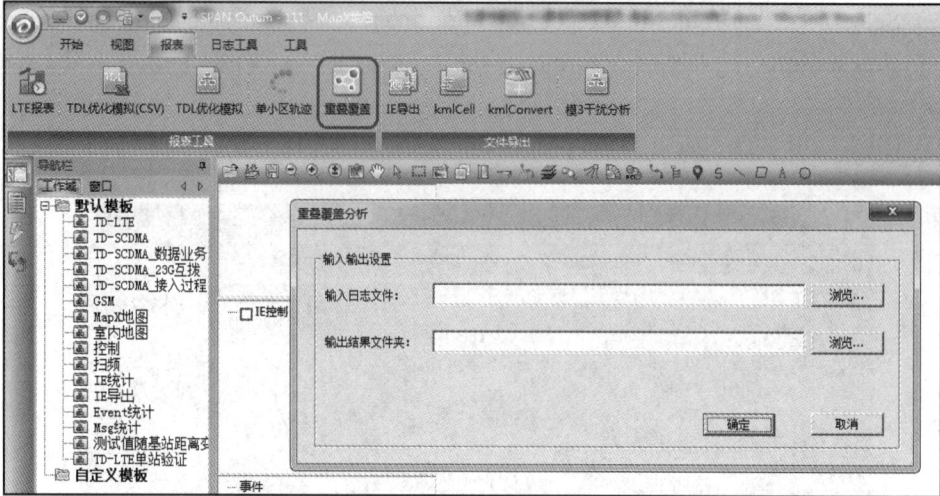

Fig. 3-6-11　Overlapping coverage

6. Report statistics

Report statistics function supports single station report and area report. Operation steps of single station report generation are as follows.

(1) Enter "background analysis" mode and click "LTE report" under the "report" branch, as shown in Fig. 3-6-12.

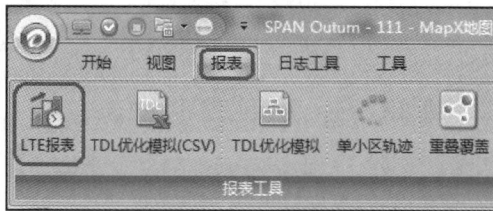

Fig. 3-6-12　LTE report

After the report setting wizard pops up, select a single station verification report and click "next", as shown in Fig. 3-6-13.

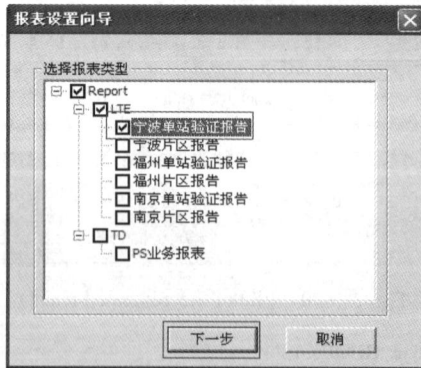

Fig. 3-6-13　Single station verification report

(2) According to the wizard prompt, click "browse", select the input log folder, and create two subfolders in this folder, with the names "single station authentication uplink service" and

"single station authentication downlink service", as shown in Fig. 3-6-14.

Fig. 3-6-14　Select the input log folder

(3) Place the LGL logs that need to generate reports in the folders corresponding to the upper and lower businesses. At present, Outum 8.10 supports the report generation function by multi-log consolidation, as well as the report generation function of up and down multiple logs simultaneously. (It should be noted that the Outum 8.10 report function does not support Chinese path currently.)

(4) According to the wizard prompt, click "browse" and select the output log folder.

(5) According to the wizard prompt, click "browse", select the cell configuration information table, and click "OK" to generate the report, as shown in Fig. 3-6-15.

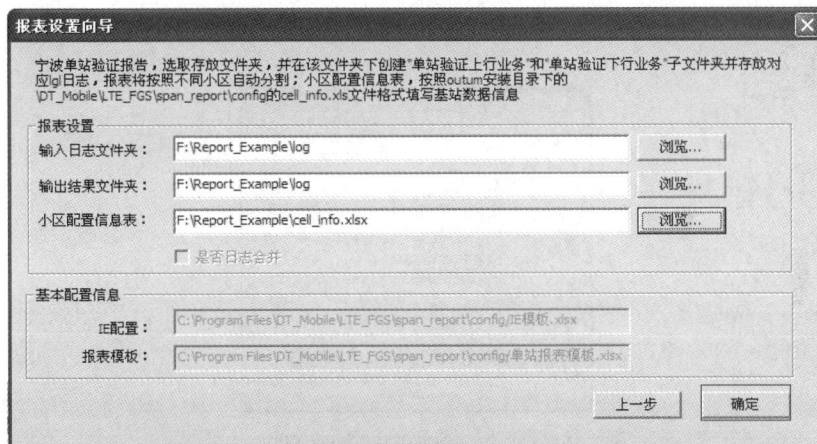

Fig. 3-6-15　Select the cell configuration information table

See Table 3-6-1 for the format of configuration information (or refer to "cell_info.xlsx" in the config folder of "span_report" in the installation directory).

Table 3-6-1　Format of configuration information

SiteName (Column fixation)	CellName (Column fixation)	PCI (Column fixation)	Frequency_DL (Column fixation)	Latitude (Column fixation)	Longitude (Column fixation)
NBJD 金融中心 (Financial center)FHTL	NBJD 金融中心 (Financial center)FHTL-0	1	1890	29.87067	121.61206
NBJD 金融中心 (Financial center)FHTL	NBJD 金融中心 (Financial center)FHTL-1	2	1890	29.87067	121.61206

It should be noted that the column position of cell table format is fixed. It is not recommended to change the column position of table.

(6) Click "OK", and the report generation result is shown in Fig. 3-6-16.

Fig. 3-6-16   Report generation result

The operation steps of the generation of area report are as follows.

(1) Enter "background analysis" mode and click "LTE report" under the "report" branch.

(2) After the report setting wizard pops up, select the area report and click "next", as shown in Fig. 3-6-17.

Fig. 3-6-17   Select the area report

(3) According to the wizard prompt, click "browse" and select the input log folder, as shown in Fig. 3-6-18.

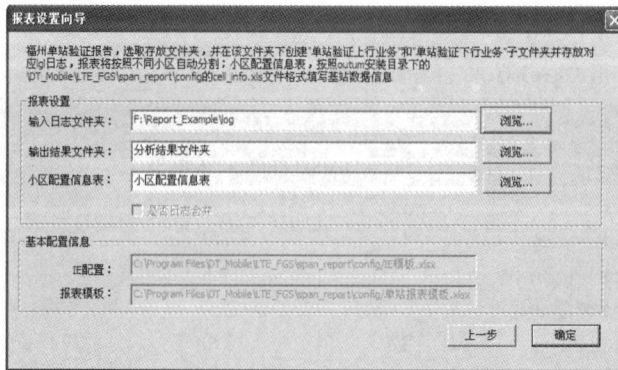

Fig. 3-6-18   Select the input log folder

It should be noted that the Outum 8.10 report does not support Chinese path currently. The later version plans to fix this problem.

(4) According to the wizard prompt, click "browse" and select the output log folder, and decide whether the generated report needs to consolidate the data exported by multiple logs.

(5) Click "OK", and the generation result of area report is shown in Fig. 3-6-19.

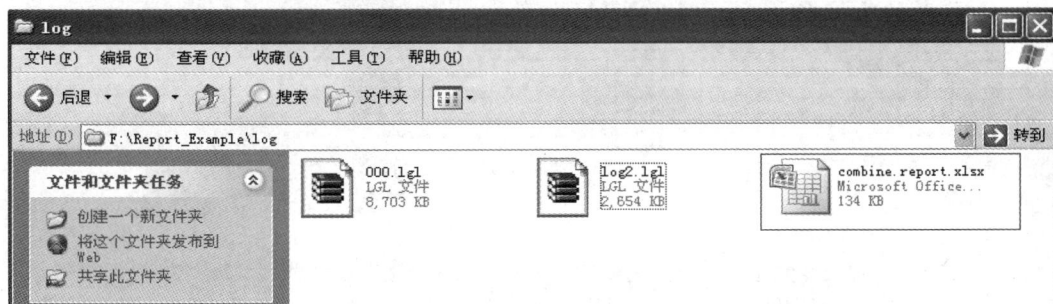

Fig. 3-6-19　Generation result of area report

7. IE export

Outum 8.10 supports IE data export, as shown in Fig. 3-6-20.

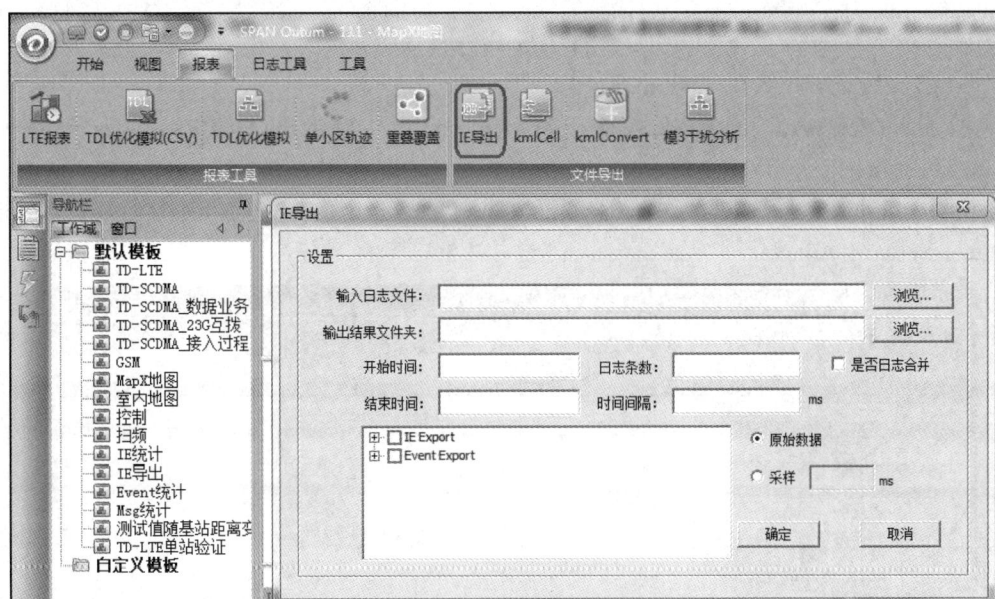

Fig. 3-6-20　IE data export

Task 7　Application Example

The application examples of Outum 8.10 are as follows.

1. Hisi FTP download

Below are the operation steps for Hisi FTP download.

(1) Start the software and create a new project, as shown in Fig. 3-7-1.

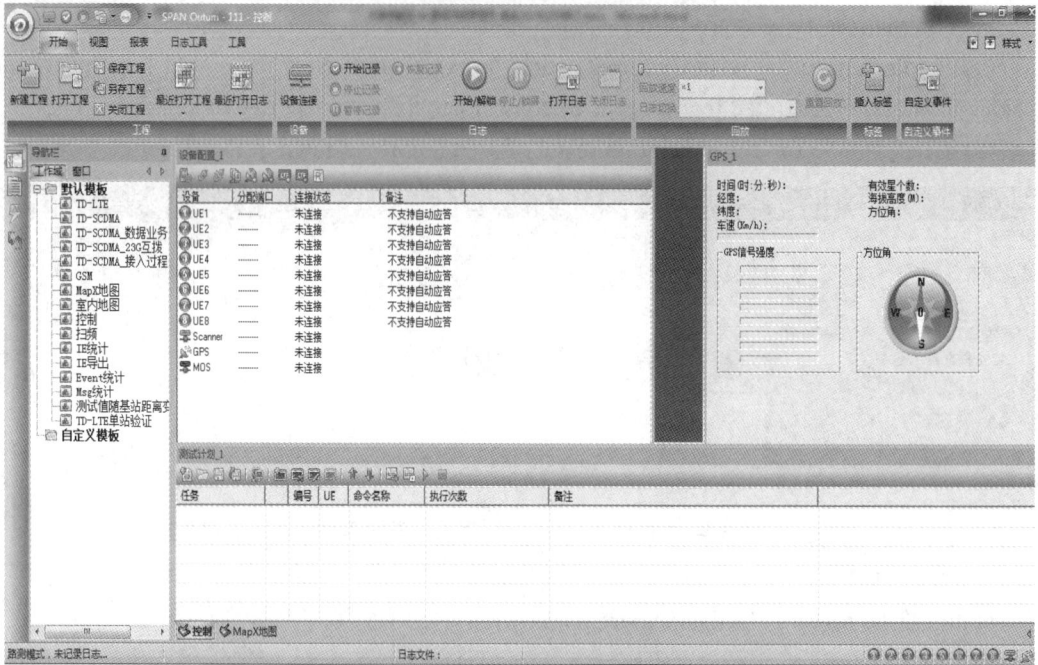

Fig. 3-7-1　Create a new project

(2) Configure the work domain.

(3) Have access to Hisi E398, GPS and other terminals. After the Hisi E398 is inserted into the PC USB interface, in order to ensure normal use, check whether there is an inserted terminal in "my computer" → "device management" → "network adapter", "HUAWEI Mobile Connect-Network Adapter #3", as shown in Fig. 3-7-2.

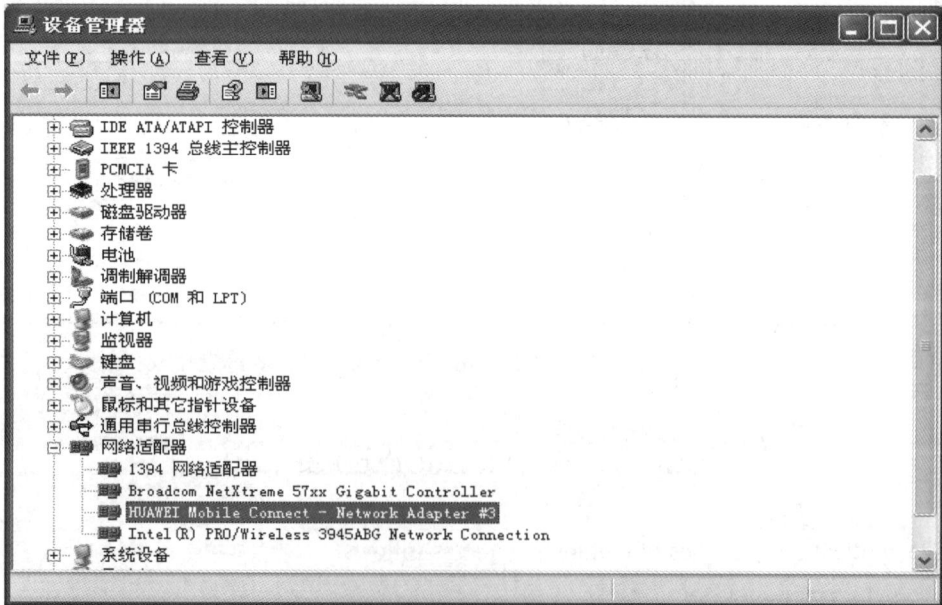

Fig. 3-7-2　Device management

Then connect Hisi E398 in the device configuration of the software, as shown in Fig. 3-7-3.

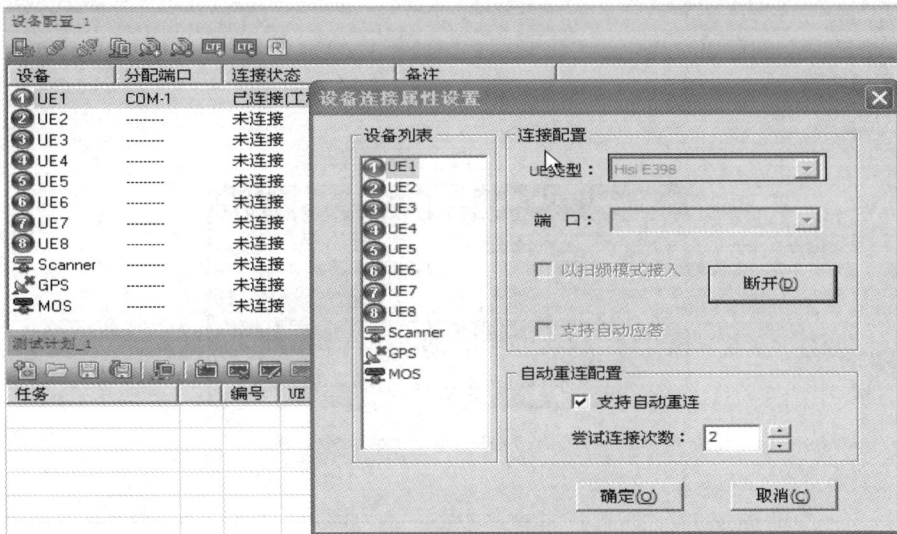

Fig. 3-7-3   Connect the terminal

When you connect the device, the "Hisi UE Agent" icon appears in the tool bar in the lower right corner of the computer. Double click the icon to open a window displaying the connection status of the connected UE, as shown in Fig. 3-7-4.

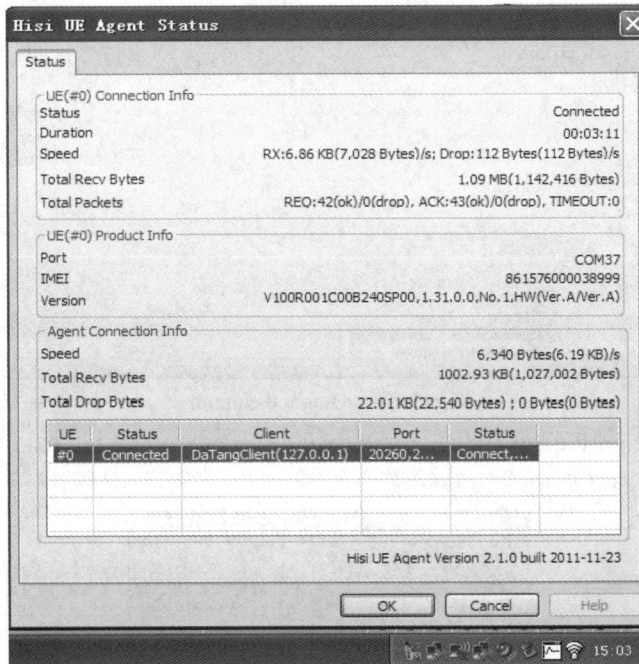

Fig. 3-7-4   Connection status of the connected UE

It should be noted that the Hisi terminal is normally connected only when both statuses shown in the above figure are connected.

(4) Import LTE project parameters. Click "BTS import" in the map domain tool bar to import LTE BTS table, as shown in Fig. 3-7-5.

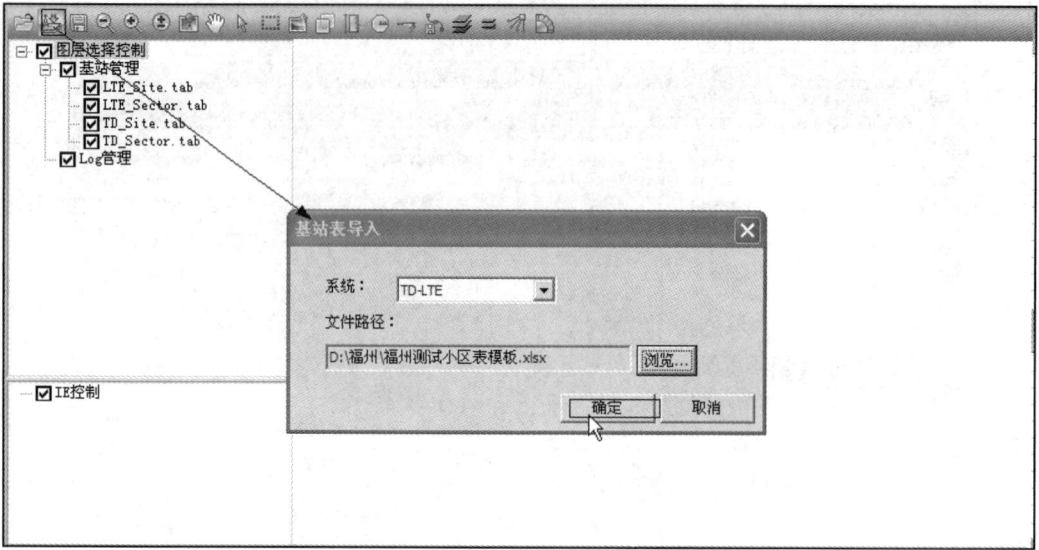

Fig. 3-7-5   BTS import

And drag the graph column (RSRP, FTP) into the IE navigation bar, as shown in Fig. 3-7-6.

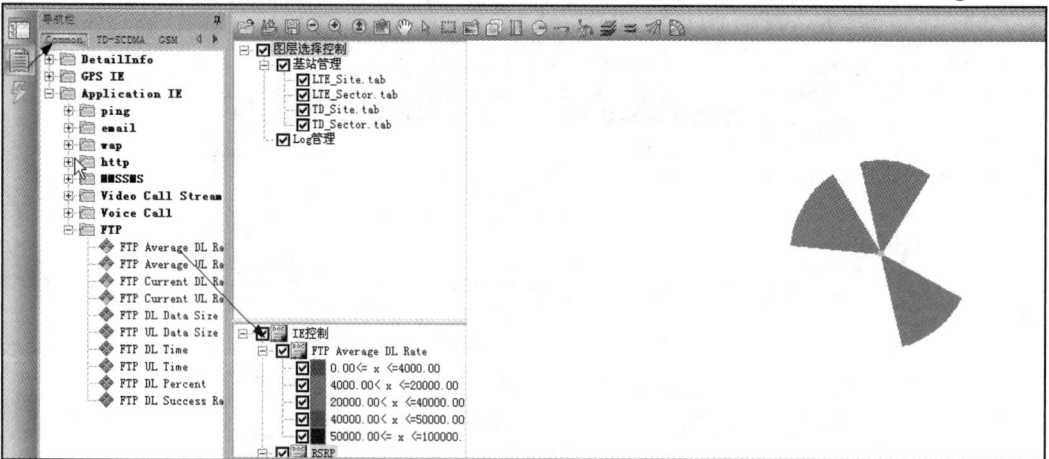

Fig. 3-7-6   Drag the column

(5) Start to record the log and enter the road test mode. Enter the road test and record the log, as shown in Fig. 3-7-7.

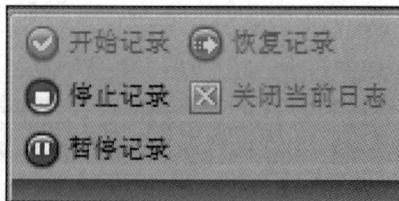

Fig. 3-7-7   Record the log

(6) Configure the third-party software too. Download the third-party FTP download tools such as 8UFtp or CuteFTP, and set the FTP server address, user name, password, port and other information, as shown in Fig. 3-7-8.

Fig. 3-7-8　Configure the third-party software tools

2. FTP download of United Microchip terminal

Below are the operation steps for download of United Microchip terminal.

(1) Refer to Hisi FTP download step (1).

(2) Refer to Hisi FTP download step (2).

(3) Have access to LC5160 and GPS terminals. Insert LC5160 into PC USB, and then check whether the terminals are correctly identified in the lower right corner of the computer. You can also check whether the port driver works normally in the network adapter managed by the computer.

Then connect the Untied Microchip terminal in the equipment configuration of the software. If the "UE type" is selected as "5160", the port will be identified automatically. If it is not correct, connect the terminal according to the given port, as shown in Fig. 3-7-9.

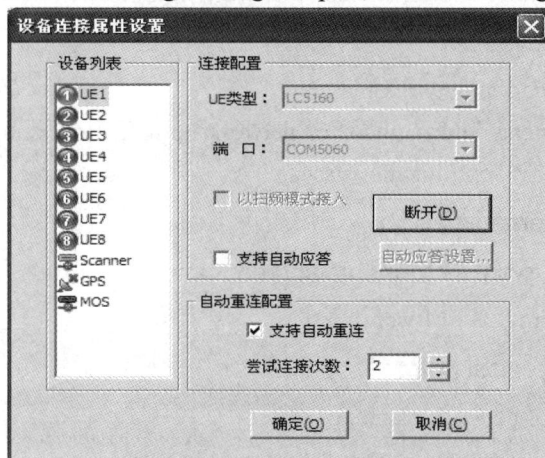

Fig. 3-7-9　Device connection property settings

(4) Refer to Hisi FTP download. step(4).

(5) Refer to Hisi FTP download step(5).

(6) Download third-party software tool. Download the third-party FTP download tools such as 8UFtp or CuteFTP, and set the FTP server address, user name, password and port, as shown in Fig. 3-7-10.

Fig. 3-7-10　Third-party FTP download tools

Training Module 4　Network Planning and Optimization

【Brief Description】

This part mainly introduces network planning, network optimization, network optimization case and high level signaling study.

【Training Elements】

1. Knowledge objectives

(1) The process and main contents of network planning.

(2) The purpose, contents and methods of network optimization.

2. Capacity objectives

(1) Master the method of network planning.

(2) Master the contents and methods of network optimization, and improve the ability to make an analysis of network optimization cases.

【Training Requirements】

1. Preparation of tools, instruments and equipment

Computer and Outum 8.10 software.

2. Knowledge appraisal point

(1) Main contents of network planning.

(2) Key parameters of network optimization.

(3) Common methods of network optimization.

3. Skill appraisal point

(1) Master the key parameters of network optimization.

(2) Master the network optimization analysis method.

Task 1　Network Planning

1. Definition of network planning

Planning is basically defined as the estimation of and decision on the development direction, development objectives and main development steps of a certain enterprise in the future.

Mobile communication network planning is to estimate and determine the future development direction, objectives and steps of the telecommunication industry.Planning needs to be considered for a long time, with emphasis on the study of development conditions, directions and objectives. The planning process is a decision-making process for a series of major issues in the development of the cause.

2. Network planning process

The basic process of network planning is shown in Fig. 4-1-1.

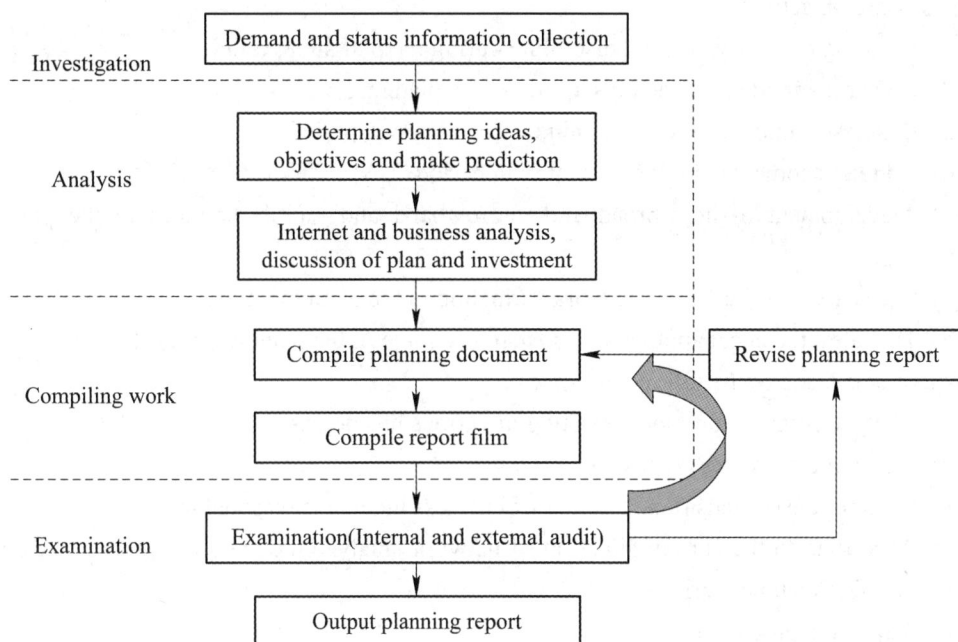

Fig. 4-1-1　Basic process of network planning

1) Investigation

(1) Local urban population and rural population.

(2) Local GDP, per capita GDP, consumption level of residents, communication expenditure, per capita disposable income, etc.

(3) Business data, revenue data, market competition strategy and tariff status of local operators.

(4) List of investments and construction projects of local operators in each year (at least 3 years).

(5) Local operators' projects under construction in the current year, quotation information of various professional products.

(6) Network construction plan and project list submitted by local operators.

(7) Equipment statistics and topological structure diagram of all specialties of local operators, such as exchange, transmission, data, synchronization, signaling, power supply, machine room, wireless, optical fiber line, pipeline, etc.

(8) Local operators' general ideas of problems of local network construction and development.

(9) Guidance documents for local network planning.

(10) Documents and faxes issued for internal tasks of local network planning, and written or electronic documents of planning drawings and technical proposals provided by equipment manufacturers.

2) Analysis

Below are objectives.

(1) Planning principles of each discipline during the planning period.

(2) Construction ideas of each discipline in the planning period.

(3) Business volume of each discipline in the planning period.

(4) Annual income targets for the planning period.

(5) Development of new broadband, narrowband and mobile services in the planning period.

(6) The scale of professional network formation in the planning period.

(7) The latest technical application of each discipline in the planning period.

Analysis is made of the following aspects.

(1) Analyze common problems existing in various businesses.

(2) Whether the business types are various.

(3) Whether the composition structure of business income is reasonable.

(4) According to the current situation of network, analyze the common problems existing in various professional networks.

(5) Safety and protection.

(6) Technology update.

(7) Equipment back from network.

(8) Smooth upgrade and expansion.

Below are scheme and investment.

(1) Networking principles.

(2) Networking ideas.

(3) Network structure.

(4) Network proposal, evolution description and attached drawings for each year.

(5) The scale of the planned network in the form of tables.

(6) According to the current quotation information of foreign and domestic manufacturers, determine the planned investment. Generally speaking, the planned equipment investment should be based on the highest contract price to improve the investment quotation.

(7) Determine the type of project investment, such as provincial management and construction, provincial management and municipal construction, municipal management and construction, etc.

3) Compiling work

(1) Typesetting should be based on the unified document template.

(2) Unify name of the construction unit, title, text, table font and chart style.

(3) Unify the drawing tools and styles of current situation map and construction scheme.

(4) Each link is organically linked to avoid independence, no correlation, or inconsistent data.

(5) Film scheduling adheres to the principle of 7 * 7 to avoid direct copying of too many words in the text.

(6) The film is mainly graphic, supplemented by text description.

(7) The color of each film is less than 4 kinds, not too fancy.

(8) Pay attention to font size and audience visibility.

4) Examination

(1) Unified joint examination.

(2) Listen to different opinions.

(3) Prediction accuracy.

(4) Rationality of the construction plan.

(5) Accuracy of investment.

3. Main contents of wireless network planning

According to the basic process of network planning, there are several aspects of mobile wireless communication network planning, as shown in Fig. 4-1-2.

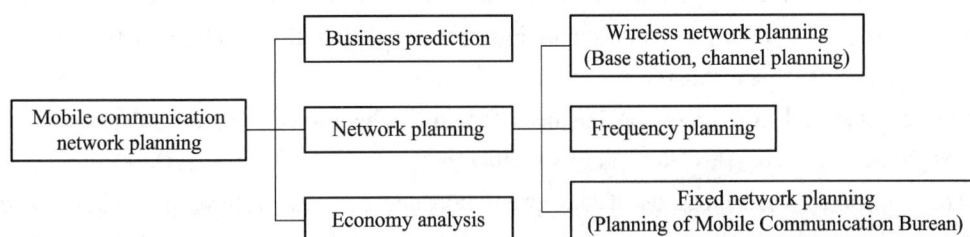

Fig. 4-1-2　Mobile wireless communication network planning

In network planning, wireless network planning is very important. The contents of wireless network planning should include determining the structure and various parameters of cellular network, the number and location of base stations, the service radius of wireless cell, channel number and capacity and the system, height and effective radiation power of antenna, etc.

4. Communication service prediction

Communication service prediction serves as the basis of wireless network planning and design. Investigation and prediction should be made according to the cycle and scale of network planning, so as to provide the necessary basis for the scientific and practical planning and design.

The prediction of communication service mainly includes three parts, user prediction, business volume prediction of each office and inter-office traffic prediction. User forecast is to forecast the number, type and distribution of users. Business volume forecast of each office is to forecast telephone business or other business volume of each exchange office. Inter-office traffic forecast is to forecast the traffic and flow direction of local or long-distance inter-office traffic or other services.

The methods of communication service prediction are as follows.

(1) According to of the forecast period, prediction can be divided into short-term forecast, medium-term forecast and long-term forecast.

① Short-term forecast: generally 2-5 years, used to guide daily operation and implement short-term plan or engineering design, etc.

② Medium-term forecast: about 10 years, which accomplish the functions of both short-term forecast and long-term forecast.

③ Long-term forecast: about 10-20 years or even longer, used for the planning, site selection, final capacity determination, pipeline and optical cable design.

(2) According to of the attribute of prediction result, it can be divided into qualitative prediction and quantitative prediction.

(3) According to of the prediction range, it can be divided into macro prediction and micro prediction.

(4) According to the nature of prediction, it can be divided into judgment prediction, historical data extension prediction and causal prediction.

(5) According to the prediction area, it can be divided into long-distance communication service prediction, local network communication service prediction and cell prediction.

(6) According to the business nature of the forecast, it can be divided into telephone business forecast and non-telephone business forecast. The forecast of non-telephone service includes public telegram, telex, fax and other data service.

The commonly used methods of user prediction and business volume prediction are time series analysis method, multi-factor correlation regression prediction method, elastic coefficient method, qualitative prediction method and penetration target method.

(1) Time series analysis. Time series is the statistical data that arranges the historical development status of the predicted object in chronological order. Based on the assumption that the future development trend of the prediction object is consistent with the past development trend and that the influence of external factors are not considered, the time series analysis method finds out the functional relationship between these statistical data and time, and selects the corresponding mathematical model for prediction. The commonly used methods are linear regression, exponential curve regression and geometric average growth rate.

(2) Multivariate correlation regression prediction. In the study of multi-user forecasting analysis, it is sometimes found that the demand for telephone (dependent variable) does not change completely with time (independent variable), and that other factors (several independent variables) have an impact on it, such as GDP, per capita economic income, average household

income, etc. Multi-factor correlation regression prediction method firstly studies various variables according to historical data, finds out the correlation and correlation degree between communication services and various factors, then uses appropriate methods to get the fitting curve, establishes the mathematical model of regression prediction, and finally makes forecast. The two kinds of prediction models for multivariate correlation regression are linear and nonlinear.

(3) Elastic coefficient method. According to the relationship between the development of telephone and the development of national economy, the elasticity index is introduced. The elasticity coefficient indicates how the growth rate of telephone communication demand adapts to the economic development.

(4) Qualitative prediction method. Qualitative prediction mainly analyzes and forecasts the future development of communication services through investigation and research. The survey objects are mainly the masses and experts in relevant fields. The survey of the masses can be conducted by means of questionnaires while the survey of experts can be conducted by means of meetings. As the preliminary work of quantitative prediction, qualitative prediction method is simple and easy, and can gather the opinions of all aspects. But this method is easy to be influenced by human factors.

(5) Penetration target method. The penetration target method is to determine the growth mode or the target level to be achieved in the expected year by considering the local economic development and communication development status, considering the telephone development level of the whole country and the whole province in the same period, and taking into account the level of the local situation in the country. Different methods can be used to predict the target level of different regions.

The main steps of communication service prediction are as follows.

(1) In order to study and predict the objective law of the development of communication business scientifically, we must first determine the prediction object, investigate and collect the development data and the various factors that may affect the prediction object, and carefully organize the resources to lay a good foundation for the prediction work.

(2) Forecast and analyze the acquired data, find out the past development law of the prediction object, and select the available prediction methods. Whether the prediction method is appropriate or not has a great influence on the prediction results.

(3) Establish a mathematical model, verify the rationality of the model, and obtain a certain reference value through specific calculation. The results obtained by this method are more scientific and simpler than those obtained by non-analytical method.

(4) The above prediction results are comprehensively analyzed, judged and evaluated, and necessary adjustments and corrections are made in certain situations to determine the final prediction results taken as the basis of communication network planning and design.

The main problems that should be paid attentions to in business forecast are as follows.

(1) When collecting historical data, we should pay attention to the statistical background

and caliber of various historical data in different periods.

(2) In order to improve the accuracy of prediction, two or more prediction methods are used. If the prediction results are similar, it is considered that the prediction results are feasible. Otherwise, it is necessary to find out the causes and select the more reasonable results.

(3) The forecast results need to be tracked, observed and corrected regularly to ensure the long-term business forecasting work, and the accuracy of business forecasting results is constantly improved.

Task 2   Network Optimization

1. LTE network optimization principle

The general steps of LTE network optimization are as follows: firstly optimize the coverage, then optimize the business performance, and finally optimize the whole network. The principles of overall network optimization contain the following four aspects.

1) Best system coverage

Coverage is an important part of optimization. In the coverage area of the system, by adjusting the antenna, power and other means, the signals at the most places can meet the requirements of the lowest level required by the service, and the limited power can be used to the best to achieve the optimal coverage, so as to reduce the user's inability to access the network, drop calls, handover failure and so on caused by the weak coverage of the system.

During the construction period of the project, the base station location, antenna parameter setting and transmission power setting can be reasonably planned according to the wireless environment. In the subsequent network optimization, the antenna parameter and power setting can be further adjusted according to the actual test situation, so as to optimize the network coverage.

When planning TD-LTE coverage, we can specify a rate target for the edge users, that is, at the edge of the coverage area, the data service of users need to meet the requirements of a specific rate, such as 64 kb/s or 128 kb/s, and even according to the business needs of some scenarios, we can propose a rate target of 512 kb/s or 1 Mb/s. As long as the actual peak rate of TD-LTE system is not exceeded, TD-LTE system can meet the different service rate target requirements of users through the allocation and configuration of system resources.

(1) Judgement of LTE system coverage. The coverage of the network, weak coverage area and over coverage area can be determined by frequency sweep and road test software. Weak coverage area means that the RSRP at the edge of the planned cell is less than −110 dBm; over coverage area means that the RSRP at the edge of the planned cell is higher than −90 dBm.

(2) Antenna parameter adjustment. Adjusting the antenna parameters can effectively solve most of the coverage problems in the network. The influence of the antenna on the network mainly includes performance parameters and engineering parameters. Antenna performance

parameters include antenna gain, antenna polarization mode, and antenna beam width. Antenna engineering parameters include antenna height, antenna dip angle, and antenna azimuth.

Generally, the appropriate antenna has been selected based on the networking requirements during the network planning and design. The performance parameters of the antenna are not suggested to be adjusted, and only when the later coverage cannot meet the requirements, and when the base station cannot be added, and when the normal network optimization means cannot solve the problem, can the antenna be replaced, such as selecting the antenna with higher gain to increase the network coverage. Therefore, in the network optimization, the antenna adjustment is mainly achieved through the adjustment of the antenna hanging height, down angle, azimuth and other engineering parameters.For example, weak coverage and over coverage are mostly solved by adjusting the pitch angle and azimuth angle of the antenna. Weak coverage can be improved by reducing the pitch angle, while over coverage can be improved by increasing the pitch angle.

(3) Antenna parameter adjustment method. In the process of single station and cluster optimization, it is necessary to ensure that the antenna parameters of each base station are verified on site. In the process of continuous optimization, the antenna parameters should be adjusted, and the data of the base station should be updated. At the same time, with the opening of new stations, the rationality of coverage still needs to be comprehensively evaluated and optimized.

2) Reasonable neighborhood optimization

Too many neighbor cells affect the measurement performance of the terminal, easily leading to inaccurate measurement of the terminal, resulting in untimely handover, false handover and slow reselection, etc. Too few neighbor cells also cause false handover, island effect, etc. The information error of neighbor cell directly affect the normal switching of the network. Three kinds of phenomena have a negative impact on the connection, drop and handoff indexes of the network. Therefore, in order to ensure stable network performance, we need to plan the neighbor cell.

(1) LTE neighborhood planning principles. Good neighborhood planning can make cell phones in the service boundary of the cell switch to the neighborhood with the best signal in time, so as to ensure the call quality and the performance of the whole network. It is the basis of good neighborhood planning to formulate the principles of neighborhood planning reasonably.

TD-LTE is basically the same as 3G neighborhood planning. The coverage of each community, station spacing, azimuth and other factors should be taken into consideration in planning. TD-LTE neighborhood relationship configuration should follow the following principles as much as possible.

① Distance principle: Areas directly adjacent to each other in geographical location are generally regarded as the neighbor cells.

② Strength principle: On the premise of optimizing the network, when the signal strength reaches the required threshold, it is necessary to consider the configuration of neighbor cell.

③ Overlapping coverage principle: The overlapping coverage area of the cell and the adjacent cell should be considered.

④ Mutual inclusion principle: Neighbor cells should be adjacent to each other, that is, sector A carrier frequency takes B as the neighbor cell, and sector B also takes A as the neighbor cell.

⑤ In some special cases, it may be necessary to configure one-way neighborhood.

(2) The principles of setting neighbor cell inside and outside the system are given below.

① Setting principle of neighbor cell in the macro station system: add all cells of the station as neighbor cells to each other; add the first circle of cells as neighbor cells; add the second circle of positive cells as neighbor cells (to be judged according to the density of surrounding station site and the distance between stations); the number of neighbor cell in the macro station is recommended to be controlled at about 8; the maximum number of neighbor cell allowed to be configured in a single cell is 32 (inside and outside the system).

② Setting principle of neighbor cell in the room subsystem: add cells with overlapping areas as neighbor cells (such as areas between elevator and each floor); add the low level cells and macro station cells as neighbor cells to ensure coverage continuity; if the macro station signal at the window side is very strong at the high-rising buildings, the macro station cells can be added to the one-way neighbor cells of the room subsystem.

③ Setting principle of neighbor cell of different systems: in addition to the internal neighbor cell planning of TD-LTE system, the neighbor cell planning between TD-LTE, TD-SCDMA, GSM and other different systems should be done well. At present, LTE mainly covers hot spots, and there are blind areas for coverage. Adding neighbor area of different systems can ensure business continuity. When setting neighbor area of different systems, TDS neighbor area is generally preferred, followed by GSM 900 neighbor area.

④ Setting principle of different systems neighborhood of macro station: add the same direction TDS/GSM cell at the same station address as the neighborhood; add the cells opposite to TDS/GSM as the neighborhood to make up for the blind coverage area; for LTE macro station at the edge of the planning area, consider adding the corresponding TDS/GSM cell as the neighborhood to ensure the business continuity; the number of different systems neighborhood of macro station is suggested to be controlled around 3.

⑤ Setting principle of neighbor area of room differentiation systems: it is recommended not to add room of different systems as neighbor area, unless it is in high traffic guarantee point, and it is suggested to add neighbor area with the same coverage and different systems to achieve load balance effect; it is recommended not to add macro station of different systems as neighbor area, except for isolated room, and to add surrounding TDS / GSM cell as neighbor area to cover blind area and ensure business continuity.

3) System interference minimization

Generally, there are two types of system interference. One is the interference caused in the system, such as improper parameter configuration, GPS deviation, RRU abnormal operation, etc. The other is the interference outside the system. These two kinds of interference directly affect the network quality.

By adjusting the power parameters, power control parameters and algorithm parameters of various services, the interference in the system is minimized as much as possible; by troubleshooting and positioning external interference, the interference outside the system is minimized as much as possible.

(1) Interference within the system. LTE has six channel bandwidth configurations, of which 5M, 10M, 15M and 20M are used as configuration options in the equipment specification. The advantages of large system bandwidth configuration are obvious, which can not only obtain higher peak rate, but also obtain more transmission resource blocks, so it is necessary to consider selecting the same frequency networking mode.

Compared with the network of different frequency, the most obvious advantage of the same frequency network is that it can use the frequency resources with higher efficiency. However, the interference between the cells causes the deterioration in the carrier to interference ratio environment of the cells, resulting in the contraction of LTE coverage, the decrease of edge user rate, and the incorrect reception of control signaling. For this reason, ICIC, power control, beam shaping and IRC can be used to effectively solve the problem of co-frequency interference in the system. In addition, through the GPS deviation detection tool and the operation and maintenance management platform (OMC) of the network element equipment, the running status and alarm of the network element equipment can be monitored in real time. Once the operation of the network element is abnormal, the operation and maintenance personnel can be notified for troubleshooting in the first place to ensure that the interference in the system caused by the failure of the network element is minimized.

(2) Interference outside the system. For the interference caused by external system, once it is found, the customer should be informed in time to solve it. In the case that the interference source cannot be determined, we can first check one by one by shutting down 1-2 circles of stations near the interfered base station in the process of initial network optimization. The external interference can be located by using the octave antenna to select the test position, antenna direction and polarization direction. The process may be long, which requires careful and patient troubleshooting by the optimization personnel.

4) Uniform and reasonable base station load

By adjusting the coverage of the base station, the load of the base station can be controlled reasonably to become as uniform as possible.

2. LTE network optimization process

The LTE network optimization process is shown in Fig. 4-2-1.

```
                          ┌─────────────┐
                          │    Start    │
                          └─────────────┘
                                 │
                          ┌─────────────────────┐
                          │ Optimization preparation │
                          └─────────────────────┘
                                 │
                          ┌─────────────────────┐
                          │ Parameter verification │
                          └─────────────────────┘
                                 │
                   ┌──────────────────────┐         ┌──────────────────────┐
                   │  Cluster optimization │────────│  Coverage optimization │
                   └──────────────────────┘         └──────────────────────┘
  ┌──────────────┐ ┌──────────────────────┐         ┌──────────────────────┐
  │ Phased output │ │ Regional optimization │        │  Business optimization │
  │ optimization  │─│                       │────────│                        │
  │   report      │ └──────────────────────┘         └──────────────────────┘
  └──────────────┘ ┌──────────────────────┐         ┌──────────────────────┐
                   │  Boundary optimization │         │    Enter next phase   │
                   └──────────────────────┘         └──────────────────────┘
                   ┌──────────────────────────┐
                   │ Whole network optimization │
                   └──────────────────────────┘
                                 │
                          ┌──────────────────┐
                          │  Accedtance check │
                          └──────────────────┘
```

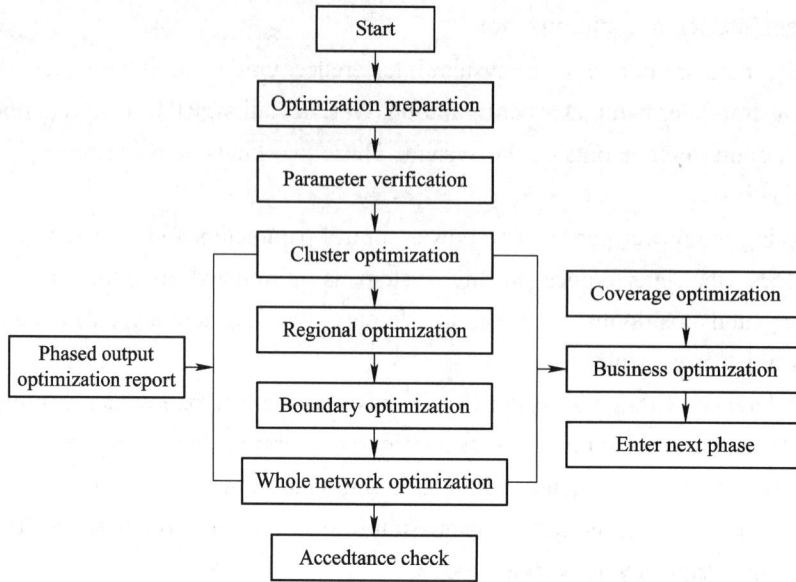

Fig. 4-2-1　LTE network optimization process

1) Optimization preparation

Before the optimization, the following preparations should be made.

(1) Base station information table: base station name, number, MCC, MNC, TAC, latitude and longitude, antenna height, azimuth, dip angle, transmission power, center frequency point, system bandwidth, PCI, ICIC, PRACH, etc.

(2) Base station opening information table, alarm information table.

(3) Map: Mapinfo electronic map of network coverage area.

(4) Drive test software: software and corresponding licences.

(5) Test terminal: test terminal matched with road test software.

(6) Test vehicle: the test vehicle according to the specific arrangement of network optimization work.

(7) Power supply: on-board power supply or UPS power supply.

2) Parameter verification

Before network optimization, first check the key parameters of the site information in the area to be optimized, and confirm whether the cell configuration parameters are consistent with the planning results. If not, submit them to the relevant personnel in time for modification.

When a station is opened, a unified start-up template can be set. Some parameters involved in the start-up template are determined by the plan. The settings of each station need to be set manually to ensure consistency. Key parameters include frequency, neighbor cell, PCI, power, switching / reselection parameters, PRACH, etc.

After the parameters are verified, the traversal coverage test is carried out for a single station to figure out the coverage of each station in detail, as well as the system performance of each sector, so as to prepare for the subsequent cluster optimization.

(1) Outdoor single station optimization. Preparations are made before test.

① Site status check: Before the site test, first prepare the cell list of multiple base stations or single base station in the area to be tested, and confirm that the status of these cells are normal.

② Configuration data check: Before the site test, collect the data of the network planning configuration and other data configured in the base station database, and check whether the actual configuration data is consistent with the planning data. Before the test, the station location, TA, UARFCN, PCI, etc. of each cell of the station to be tested must be obtained.

③ Test site selection: In order to ensure that the test service is provided by the cell to be tested, when selecting the test point, select the location where the signal strength of the target cell is strong and the signal of other cells is relatively weak for cell equipment function test.

Below are verification works in IDLE mode.

① Frequency check: Check whether the frequency points of each cell are correct on the road test software.

② PCI check: Check whether the PCI of each cell is correct on the road test software.

③ TA check: Check whether the TA of each cell is correct on the road test software.

④ Cell reselection: Test and check whether the cell reselection parameters are set normally on the road test software, and reselect the cell in the station.

Below are verification works in CONNECT mode.

① Success rate of Attach activation: When the terminal randomly accesses the network and conducts Attach activation, the success rate of Attach activation needs to be recorded. Problems need to be located and solved before retesting.

② Success rate of random access: After the completion of Msg5, the random access success rate of terminal needs to be recorded. If there is any problem, it needs to be located, solved and retested.

③ Paging test: The network side issues paging instructions to check whether the terminal can smoothly enter the ACTIVE state from IDLE state.

④ Switching test: Use UE for continuous test to check whether the switching is normal.

⑤ Upload and download service test: The terminal carries out the specified FTP service for 3 minutes and records the upload and download rate.

(2) Indoor single station optimization. Preparations are made before test.

① Information preparation: Before the test, it is necessary to obtain the relevant information of the test point in advance, including the site name, location, indoor area, outdoor neighbor cell, floor plan, system design drawing, property contact person and contact information, information of the construction integrator of the test point, etc.

② Test equipment and instrument preparation: 1 set of test terminal and data line, 8 other terminals (including charger); 1 set of test software and software dog, 1 notebook computer, 2-3 data cards, SIM cards (corresponding to the number of all terminals and data cards), 1 signaling tracking analyzer (such as K1297), 1 SiteMaster, 1 battery, 1 inverter and 1 set of multi-purpose socket.

③ Preparation of testers: 2-3 testers and their contact information.

Below are coverage performance tests.

① Select the indoor test route, which should traverse the main indoor coverage area.

② According to the test route, the typical antenna points are selected, and the RSRP power of the antenna port signal at this point is tested by spectrum analyzer. The sampling quantity of each layer should not be less than 30% of the total number of layers.

③ Turn on the road test software, and conduct the test in the room at walking speed along the test route. The road tester records the received RSRP, SINR, PCI and other data.

④ Turn on the road test software, and walk around the building at a walking speed within 20 meters of the outdoor area. The test terminal locks the indoor target community for testing. The road tester records the received data such as RSRP, SINR, PCI, etc.

Below are coverage index requirements.

① Meet the relevant national environmental protection requirements, and the electromagnetic radiation value must meet the requirements of the national standard "electromagnetic radiation protection code" The maximum transmission power of indoor antenna carrier is less than 15dBm.

② RSRP>−85dBm, SINR>15 dB in 90% of indoor areas.

③ Indoor users are provided with dominant frequency by indoor coverage system, which is at least 5dB higher than the strongest outdoor signal.

④ Indoor leakage control, RSRP<−90 dBm at 20m outside the signal leakage building.

⑤ The output power of each antenna is≥65 dBm, and the PCCPCCH power of the antenna port is 0-5dBm. Considering the coverage requirements, it can reach 7dBm in some cases, and the deviation from the design value is not more than 3 dBm.

Below are system performance tests.

① The success rate of Attach activation.

② Switching test in the system.

③ Ping test.

④ Upload and download business test.

3) Cluster optimization

A cluster contains generally 20-30 sites. Before starting cluster optimization, in addition to confirming that the base station has been opened, it is also necessary to check whether there are alarms in the base station to ensure that the optimized base station works normally. Here we first introduce the concepts of coverage optimization and business optimization.

(1) Coverage optimization is the first step of engineering optimization, and also the most important and basic step, the focus of which are RSRP, RSRQ (Reference Signal Received Quality) and CINR. The main optimization method is to adjust the engineering parameters and power, as well as the relationship between neighbor cells. After covering the engineering

parameters and RS power, the engineering parameter table needs to be updated in time. In addition to the adjustment of neighbor cell list and frequency scrambling code, the adjustment measures in RF optimization stage are mainly the adjustment of engineering parameters. Most of the coverage and interference problems can be solved by adjusting the following station engineering parameters in the order of priority: ① antenna dip angle; ② antenna direction angle; ③ antenna type; ④ antenna height; ⑤ antenna location; ⑥ change station type; ⑦ station location; ⑧ new station.

(2) Business optimization refers to the optimization of all businesses required by the planning after the coverage optimization meets the index requirements. First, test the long call of each basic service, checking whether the handover success rate and data service rate meet the index requirements, and then test the short call of each basic service, checking the access success rate and call drop rate. In view of the situation where the business does not meet the index requirements, it is necessary to analyze the reasons and optimize the adjustment. After optimization and adjustment, the engineering parameter table and parameter adjustment tracking table should be updated in time.

(3) Cluster optimization data collection.The correct collection of data is the premise of doing well in the optimization work. If the data is not taken correctly, it brings difficulties in finding and solving network problems. In the RF optimization stage, we focus on the optimization of wireless signal distribution in the network. The main test methods are DT test and indoor test. Before the test, check with the maintenance engineer of the computer room whether there is any abnormality in the base station to be tested, such as shutdown, blocking, congestion, transmission alarm, etc.; judge whether there is a negative impact on the data authenticity of the test results; if so, arrange the test after removing the alarm.

① Data acquisition is mainly based on DT test. Through DT test, wireless signal data of Scanner or UE is collected for analysis of outdoor signal coverage, switching and no dominant frequency. It is important to do the following in DT test.

A. DT test passes through as many base stations as possible.

B. DT test covers main roads in network coverage area.

C. The vehicle can run at different speeds on the test route.

D. DT test is done in different propagation environments, Direct, reflection and deep fading.

E. DT test route should cross the overlapping coverage area of base station.

F. DT test adopts two UEs for long call and short call test. Long call test setting: it is recommended to keep the call for more than 1 hour. Short call test setting: it is recommended to keep the call for 90 seconds and idle for 20 seconds.

② Indoor test is mainly aimed at indoor coverage areas (such as buildings, shopping malls, subways, etc.), interior of key places (such as gymnasiums, government agencies, etc.), and other areas required by operators for signal coverage test, so as to find, analyze and solve RF problems of these places. In addition, indoor test can help to optimize the switching relationship between indoor and outdoor same frequency, different frequency or different systems.

(4) Cluster optimization data analysis.Coverage analysis is the focus of RF optimization, with emphasis an on signal distribution. Weak coverage, cross region coverage, uplink and downlink imbalance, and no primary service cell fall into the scope of coverage analysis.

① Weak coverage.Weak coverage means that the RSRP of pilot signal in coverage area is less than −100 dBm, the area like concave land, back of hillside, elevator shaft, tunnel, underground garage or basement, interior of tall buildings, etc. If the pilot signal is lower than the minimum requirement of the full coverage service, or just meets the requirement, and SINR cannot meet the minimum requirement of the full coverage service due to the increase in the same frequency interference, problems are such as difficult access and dropping of the full coverage service. If the pilot signal RSRP is lower than the coverage area in the minimum access threshold of the hand machine, the mobile phone usually cannot stay in the cell and the problem of "off grid" occurs due to failure in location update and location registration. The following countermeasures are used to resolve these problems. The coverage can be optimized by adjusting the antenna direction angle and dip angle, increasing the antenna hanging height, replacing the higher gain antenna, etc. When there are more users in the non-overlapping part of the adjacent base station coverage area or when the non-overlapping part is large, new base station should be built, or the coverage range of the surrounding base station should be increased to improve the coverage overlapping depth of the two base stations. The size of the switching area should be considered, and it's necessary to pay attention to the interference of the same adjacent frequency caused by the increase in the coverage. For the weak coverage area caused by the depression and the back of the hillside, new base stations can be added to extend the coverage. For the signal blind area inside the elevator shaft, tunnel, underground garage or basement, high building, solutions such as RRU, indoor distribution system, leakage cable, orientation antenna can be used.

② Cross region coverage. Cross region coverage generally means that the coverage area of some base stations exceeds the planned scope, forming a discontinuous dominant area in the coverage area of other base stations. For example, for some stations that are much higher than the average height of the surrounding buildings, the transmission signal can travel far along the hilly terrain or roads, forming a dominant coverage in the coverage area of other base stations, resulting in the phenomenon of "island". Therefore, when the call is connected to the "island" area far away from a base station and still served by the base station, and when the cell handover parameter is set and the cell around the "island" is not set as the adjacent cell, once the mobile station leaves the "island", drop calls immediately occur. And even if the neighbor cell is configured, it is likely that drop calls happen due to the narrow area of the "island". Such a phenomenon happens on both sides of the harbor where mutual coverage may lead to signal interference. The following countermeasures are recommended to deal with problems discussed above. For the cross-region coverage, it is necessary to avoid the antenna transmitting to the road as much as possible, or use the surrounding buildings as a shielding to reduce the cross-region coverage, but at the same time, it is necessary to avoid same frequency interference in other base stations. For high station, it is recommended to change its location, but it be may hard to achieve due to limitations from property management and equipment installation. Moreover, adjustment of the mechanical dip angle of the antenna can cause the distortion of the antenna pattern, so in most cases measures such as adjustment of pilot power or the adoption of electric dip antenna can be used to reduce the coverage of the base station to eliminate the "island" effect.

③ Imbalance between uplink and downlink.The imbalance between uplink and downlink refers to the case that in the target coverage area the downlink coverage performs well, while the uplink coverage is limited (the transmission power of UE reaches the maximum but still cannot meet the requirements of uplink BLER) or the downlink coverage is limited (the transmission power of downlink special channel code reaches the maximum but still cannot meet the requirements of downlink BLER). The unbalanced coverage of uplink and downlink is likely to lead to dropped calls. Limited uplink coverage is more frequent.The following countermeasures are recommended to deal with problems discussed above. For the uplink and downlink imbalance caused by uplink interference, it is necessary to check whether there is interference by monitoring the alarm condition of the ISCP of the base station, and then deal with the interference by referring to the relevant instructions. Other reasons may also cause the uplink and downlink imbalance: problem in the uplink and downlink gain setting of and dry amplifier equipment; problem in receiving diversity antenna of the transceiver separation system; problem in eNodeB hardware. The working state of the equipment should be checked, and measures such as replacement, isolation and local adjustment can be used to deal with such problems.

④ No leading cell signal.This kind of area refers to places where there is no dominant cell or the dominant cell is replaced too frequently, which lead to frequent switching, thus reducing the system efficiency and increasing the possibility of dropped calls.

For the area without dominant cell, the coverage of one strong signal cell (or short-range cell) should be enhanced and the coverage of other weak signal cell (or long-distance cell) should be weakened by adjusting the antenna dip angle and direction angle.

(3) Analysis of cluster optimization handover.In the RF optimization phase, the handover issues involved are mainly neighbor cell optimization and handover area control. By adjusting the RF parameters, the size and position of the switching area can be controlled, reducing call drops caused by rapid signal changes and improving the success rate of switching. There are two cases in neighborhood optimization neighborhood adding and neighborhood deleting. The impact of missing neighbor area is that the strong cell cannot join the neighbor area list, resulting in increased interference or even dropped calls. In this case, the necessary neighbor area needs to be added. Redundant neighbor area may lead to huge amount of message and the increase in unnecessary signaling overhead, and besides when the neighbor area is fully configured, necessary area cannot join it. In this case, the redundant neighbor area needs to be deleted.

① Add neighbor cell. In the RF optimization stage, we mainly focus on the situation of missing neighbor area. The methods of adding neighbor area are as follows: add neighbor area according to geographical location; add missing neighbor area through software inspection according to geographical location; add neighbor area according to road test data; add neighbor area according to Scanner data; add neighbor area according to UE test data. According to the distribution of RSRP and SINR, analyze some cases of dropped calls and call failure, and record the cells with missing neighbor area.

② Deletion of redundant neighbor cell. The deletion of redundant neighbor cell needs great caution. Once necessary neighbor cell is deleted by mistake, serious consequences such as dropped calls arise. Therefore, before deleting the neighbor cell, it is necessary to check the modification records of the neighbor cell, and to confirm that the neighbor cell to be deleted is not the neighbor cell added in the previous road test and optimization. After deleting the redundant neighbor cell, a comprehensive test, including road test and important indoor location dial test, is required to ensure that no abnormality occurs, or otherwise the data configuration needs to be changed back. In the RF optimization stage, neighbor cell may be deleted in the following cases: delete the neighbor cell relationship of cross area coverage, provided that the cross area coverage problem has been solved and no new weak coverage area has been added; delete the neighbor cell based on the network topology, which is applicable to the case that the original neighbor cell table is full, and new neighbor cell relationship needs to be added. After deletion, test shall be arranged to confirm that no more problems occur, or otherwise, we need to reselect the neighborhood to be deleted.

(4) Analysis of cluster optimization and adjustment. The analysis mainly includes the following aspects.

① Adjust according to the Scanner test data: according to the RSRP distribution of each cell in the Scanner test network, if there are overlapping coverage areas of the same cell code or Mod 3, it needs to be adjusted.

② Adjust according to UE test data: through the analysis of UE test data, we focus on the places where SINR suddenly jumps. If there is overlapping coverage of cells with the same code, SINR gets worse, BLER rises sharply, and dropped calls and other events are likely to occur.

③ Check by software: according to PCI planning law, the rationality of planning can be checked by software. It is required that the multiplexing distance of the same code is more than 3km, and if it is less than 3km, there is a problem in the planning, which needs to be readjusted. The key parameters are latitude and longitude, azimuth, and frequency point code word.

④ The control of call rate is related to the parameters of reference signal transmitting power, uplink PRACH power, cell downlink access power threshold, SRS period, etc.

⑤ Call failure caused by cell reselection: the failure is related to downlink minimum access threshold of cell selection / reselection, measurement trigger threshold of cell reselection at the same frequency, measurement trigger threshold of cell reselection between frequencies, service cell reselection delay, cell reselection time delay and other parameters.

⑥ The related wireless parameters of handover success rate are related to cell personality shift, handover time delay, handover RSRP latency, etc.

⑦ Control of dropped calls: dropped calls are caused by weak coverage and uplink interference.

4) Regional optimization

After the optimization of each cluster in the divided region, the coverage optimization and business optimization of the whole region are carried out. The optimization focuses on cluster boundary and some blind spots. Coverage optimization comes first and then is followed by business optimization. Its process is the same with that of the cluster optimization. It is suggested to form a network optimization group to optimize the boundary. During the optimization process, it's necessary to update the engineering parameter table and parameter adjustment tracking table in time, and summarize the comparison report before and after the adjustment.

5) Boundary optimization

After the optimization in the region is completed, the region boundary optimization is started. A joint optimization team is composed of network optimization engineers from neighbor cell to optimize the boundary coverage and service. When there are different manufacturers on both sides of the boundary, the engineers of the two manufacturers need to form a joint network optimization team to optimize the coverage and business of the boundary.

6) Whole network optimization

The whole network optimization refers to the overall network DT test for the whole network, the overall understanding of network coverage and business conditions, and coverage and business optimization for key roads and areas provided by customers.

3. Key parameters of LTE network optimization

1) Related parameters of coverage

(1) RSRP is defined in the protocol as the linear average of all RE power carrying RS within the measured bandwidth, see 3GPP 36.214. The measurement reference point of UE is antenna connector, and the measurement state of UE includes RRC_IDLE state and RRC_CONNECTED state within and between systems.

(2) RS-CINR Carrier to Interference plus Noise Ratio. RS-CINR is defined as the strength of RS useful signal compared with interference (or noise or interference plus noise) at the terminal.

RS-CINR indicates the quality of channel coverage. RS-CINR is related to network load. The higher the network load, the worse the RS-CINR. Therefore, the optimization objectives of RS-CINR are different under different network loads.

(3) RSRQ. The definition of RSRQ in the protocol is $N \times$ RSRP/(E-UTRA carrier RSSI),

that is, RSRQ = 10lg(N)+the RSRP － RSSI of the main service cell that UE's location receives. N refers to the number of RB in the frequency bandwidth of UE measurement system, and RSSI refers to the linear average of the power on OFDM symbols containing reference signals on the antenna port0. First, we accumulate the received power on all RE within the measurement bandwidth of each resource block, including useful signals, interference, thermal noise, etc., and then the linear average is carried out on the OFDM symbols. See 3GPP 36.214.

RSRQ is not only related to the RE power carrying RS, but also related to the RE power carrying user data and the interference of neighbor cell. Therefore, RSRQ changes with the network load and interference. The larger the network load, the greater the interference, and the smaller the RSRQ measurement value.

(4) PDCCH SINR. SINR, refers to the ratio of the strength of the received useful signal to the strength of the received interference signal (noise and interference).

The general calculation formula is PDCCH SINR = (channel receiving power of the best service cell / interference of the coverage cell channel).

PDCCH SINR indicates the quality of PDCCH channel. The demodulation threshold of PDCCH channel is defined in 3GPP 36.101.

(5) RSSI is The total received power on all the RE of an OFDM symbol in UE detection bandwidth (if it is a system bandwidth of 20 M, when there is no downlink data, it is the total received power on 200 RE, and when there is downlink data, it is the total received power on 1200 RE), including the signal of service cell and non-service cell, adjacent channel interference, system internal thermal noise, etc. That is, the total power $S + I + N$, where S is the signal power, I is the interference power and N is the noise power. It reflects the received signal strength and interference degree of the current channel.

2) Reselect related parameters

(1) Treselection. It indicates that the duration of the measurement cell with better channel quality than the current cell is not shorter than Treselection, and the UE can only initiate the same frequency cell reselection process after staying in the original cell for more than 1s.

(2) Q-RxLevMin. It refers to the minimum channel quality requirements for cell reselection with different frequencies. InterFreqRxLevMin means the minimum required reception level for cell reselection, or receiver sensitivity. This parameter is configured by Q-RxLevMin in InterFreqCarrierFreqInfo in the RRC signaling SIB (System Information Block) 5 message specified in 36.331.

According to the budget analysis of RsPower's link, it can be determined that the demodulation threshold of PDCCH (8CCE) is in the middle on the premise of the same transmission power, and the power improvement of other channels is based on PDCCH (8CCE). Therefore, the minimum required reception level of cell reselection should meet the demodulation performance of PDCCH (8CCE).

(3) Offset. It refers to the Offset value of cell reselection.

3) Switch related parameters

(1) Hysteresis: A3 event trigger lag factor. This parameter indicates the lag factor used in the entry and exit conditions of event triggered escalation, which works together with A3Offset. When the RSRP of the measurement cell is higher than that of the service cell (A3Hysteresis + A3Offset), the A3 measurement report is triggered.

(2) Time to Trigger. TTT event triggers a continuous event; a measurement report is triggered of the duration of the event criteria is met, i.e. the corresponding measurement report can only be triggered after meeting the entry or exit conditions of an event.

(3) A3Offset. It works together with the lag factor. Generally, when the channel quality of adjacent cell is better than that of the cell, the handoff occurs. Therefore, the value of this parameter should be positive.

4. Comparison between TD-SCDMA and TD-LTE network optimization

LTE is closely related to 3G, which is shown in the following aspects.

(1) In the same frequency network, the network performance is mainly limited by the interference in the system, and the idea and means of wireless planning and optimization are relatively similar.

(2) Similar coverage: TD-LTE and TD-SCDMA are close to each other in the frequency band of 2GHz, with similar coverage. At the same time, the existing coverage of 3G can be used as a reference for LTE coverage prediction.

(3) Services are close to each other: 3G services such as PS services and video services have cultivated LTE services. Meanwhile, the analysis and optimization methods of user service performance by 3G are worth learning, such as the determination of some KPI indicators.

(4) Key technologies are close to each other: 3G HSPA technology introduces AMC, HARQ, multi-antenna / smart antenna and other technologies, which can be continued in LTE.

From test experience, the coverage capacity of 2.6G TD-LTE is slightly weaker than that of TD-SCDMA (meeting certain edge rate requirements). Under the condition of common station construction, TD-LTE can obtain good outdoor continuous coverage. However, as the main production area of high-speed business, further solution for hotspot and indoor coverage are needed for the fact that some of these areas are difficult to meet the coverage requirements of high-speed output and transmission.

TD-SCDMA system adopts multi-frequency network to reduce the same frequency interference of pilot channel and broadcast channel. When planning the system, we need to do the scrambling planning to ensure that the signals of different cells are not mixed up.

TD-LTE system can effectively control CO frequency interference by adopting the mechanism of interference randomization, interference elimination and interference coordination. In LTE system, many physical channels/signals, such as PSS/SSS, PCFICH, etc., are related to

PCI. The planning of PCI is very important. If PCI planning is not good, relatively large neighbor cell interference is caused, thus reducing cell throughput and increasing transmission delay.

PCI planning should start with a study of the constraints between adjacent cells, and then conduct the plan of the whole network PCI with a reasonable planning algorithm based on the research results.

The constraints of PCI between adjacent cells: Collision-free, PCI of two adjacent cells should be different; Conflict-free, PCI of all adjacent cells cannot be the same; the remainder of PCI mode 3 of two adjacent cells should be different. Use a reasonable planning algorithm to allocate PCI for the whole network:The neighborhood relationship is calculated according to the actual network topology; according to the neighborhood relationship, PCI is allocated to all cells, and the multiplexing distance of PCI is as far as possible.

Task 3   Network Optimization Case

The network optimization cases and network optimization analysis methods are as follows.

1. Coverage analysis of TD-LTE network

According to the "network test data in Kai County" collected by Outum 8.10, the corresponding cell table is "Wanzhou County LTE parameters". After playback and analysis, the pull line diagram of UE1 coverage is shown in Fig. 4-3-1.

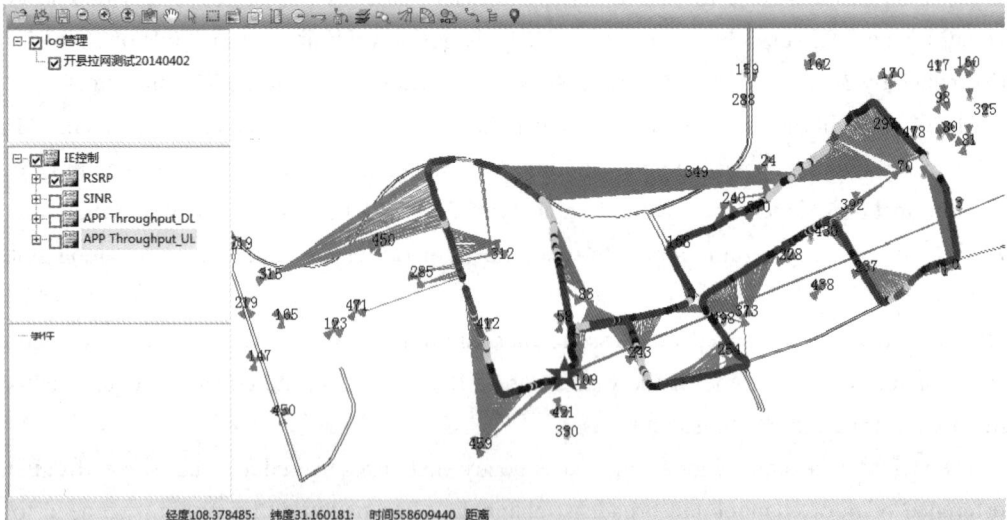

Fig. 4-3-1   Pull line diagram of UE1 coverage

Use the Outum 8.10 to output UE1 coverage interval scale diagram and histogram, as shown in Fig. 4-3-2.

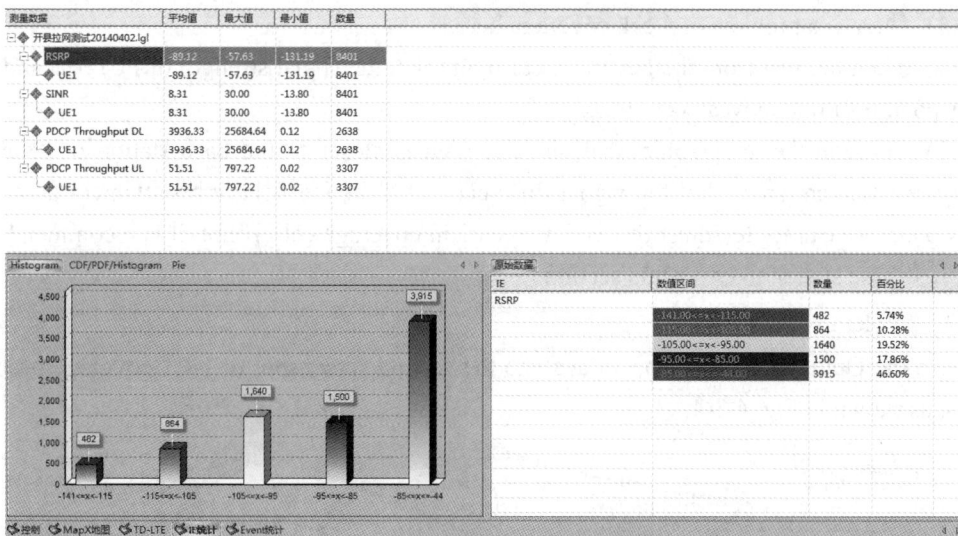

Fig. 4-3-2　UE1 coverage interval scale diagram and histogram

From the average value of overall coverage and the proportion of RSRP>−105dBm, the average value of this coverage is −89.12 dBm, and the proportion of RSRP>−105 dBm is 80.48%.

1) Problem section 1

Use the Outum 8.10 and map to analyze the first screenshot of UE1 coverage problem points, as shown in Fig. 4-3-3.

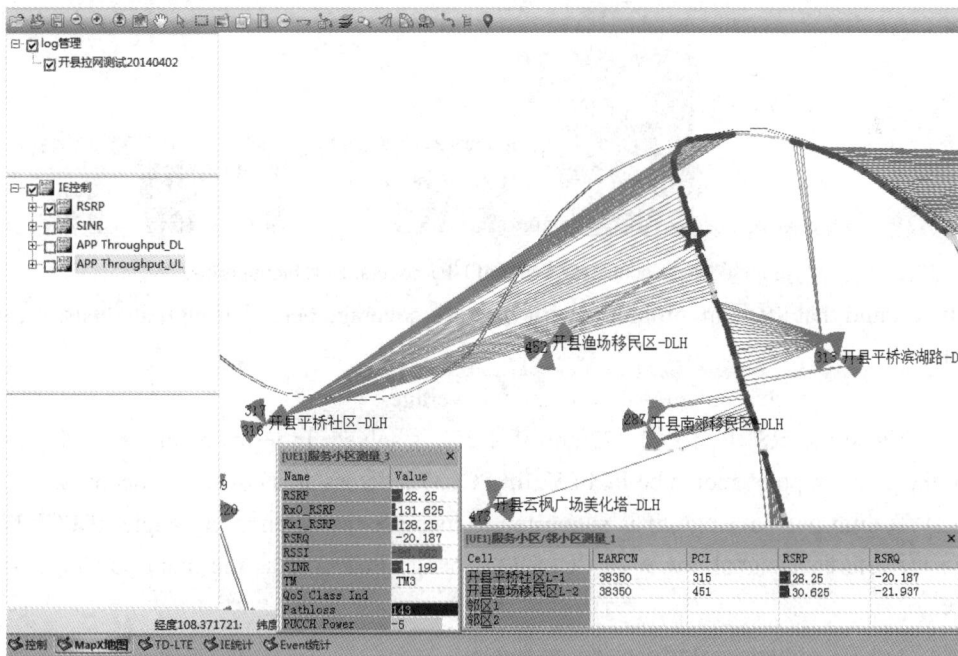

Fig. 4-3-3　First screenshot of UE1 coverage problem points

It is found that there are problems with network coverage here. Through analysis, it can be seen that:

(1) There is a problem of weak coverage here.

(2) It can be seen from the geographical figure that the signal strength of PCI 315 and PCI 451 is poor, resulting in weak coverage.

(3) Through the geographic analysis, this area is closer to the base station of "Kaixian fishery resettlement area", but the signal difference of "Kaixian fishery resettlement area L-2" cell is poor. We can try to adjust this cell as the main coverage cell. If not, it is recommended to add a new base station to cover this section.

2) Problem section 2

Use the Outum 8.10 and map to analyze the second screenshot of UE1 coverage problem points, as shown in Fig. 4-3-4.

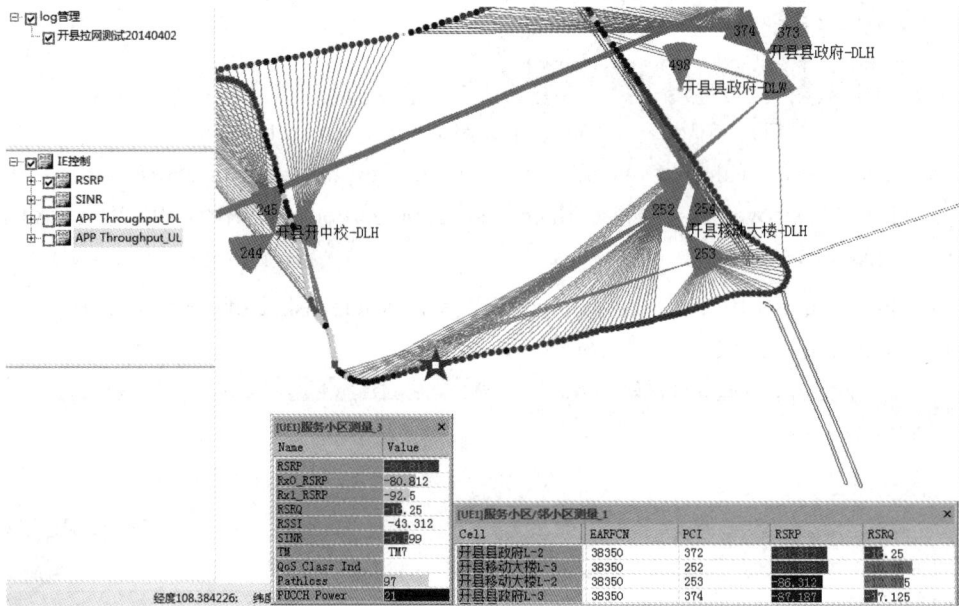

Fig. 4-3-4　Second screenshot of UE1 coverage problem points

It is found that there are problems with network coverage here. Through analysis, it can be seen that:

(1) This coverage problem is no dominant coverage.

(2) The signal RSRP of PCI 372 and PCI 252 is solved. In terms of distance, the nearest cell to the problem point should be PCI 252. In PCI 372, "cross area coverage" occours.

(3) To improve coverage, it is suggested to increase the antenna dip angle of PCI 372 or reduce the reference signal power of PCI 372. Change PCI: 252 as the main coverage of this section.

2. TD-LTE network interference analysis

In the "Lishui test data. Lgl" collected by Outum 8.10, the corresponding cell table is "Lishui LTE-OUTUM8 import information table". After playback and analysis, the UE1 SINR pull diagram is shown in Fig. 4-3-5.

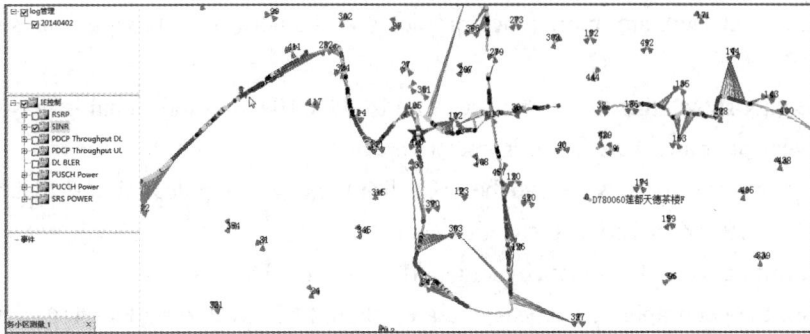

Fig. 4-3-5　UE1 SINR pull diagram

Use the Outum 8.10 to output UE1 SINR interval scale diagram and histogram, as shown in Fig. 4-3-6.

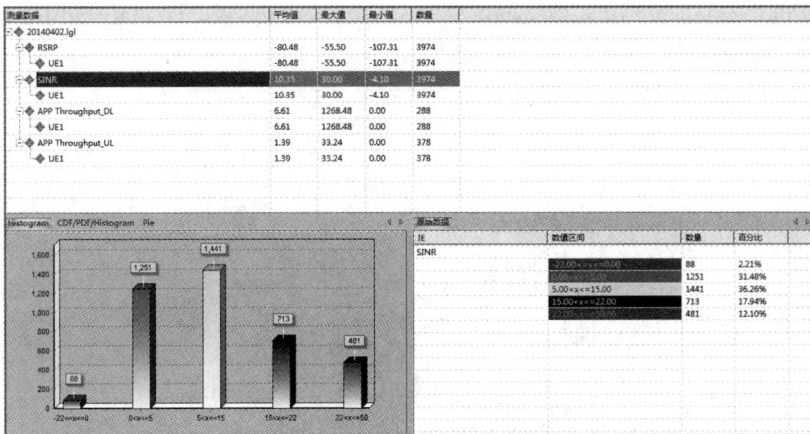

Fig. 4-3-6　UE1 SINR interval scale diagram and histogram

The average value of downlink interference and the ratio of SINR>0 are evaluated by the interval scale and histogram, of which the average value of SINR is 10.35 dB, and the ratio of SINR greater than 0 is 97.79%.

1) Problem section 1

Use the Outum 8.10 and map to analyze the first screenshot of UE1 interference problem points, as shown in Fig. 4-3-7.

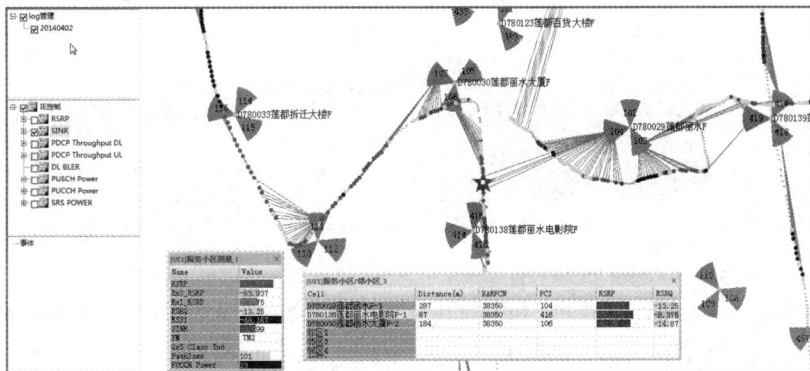

Fig. 4-3-7　First screenshot of UE1 Interference problem points

It is found that there are interference problems in the network. Through analysis, it can be seen that:

(1) The data shows that this area is covered by PCI 104, and the signal is very strong, but the SINR is very poor and there is cross area coverage.

(2) Through the analysis, it can be seen from the geographical figure that the area is covered by PCI 104, 416 and 106 together, and the signals are very strong without obvious difference, resulting in no dominant coverage, affecting the SINR value here.

(3) From the geographical analysis, the area should be covered by PCI 416, but the signal of PCI 104 is seriously over covered, so in PCI 104, we should increase the dip angle to suppress the over covered area, so as to improve the SINR value.

2) Problem section 2

The second screenshot of screenshot of UE1 interference points analyzed by Outum 8.10 and map is shown in Fig. 4-3-8.

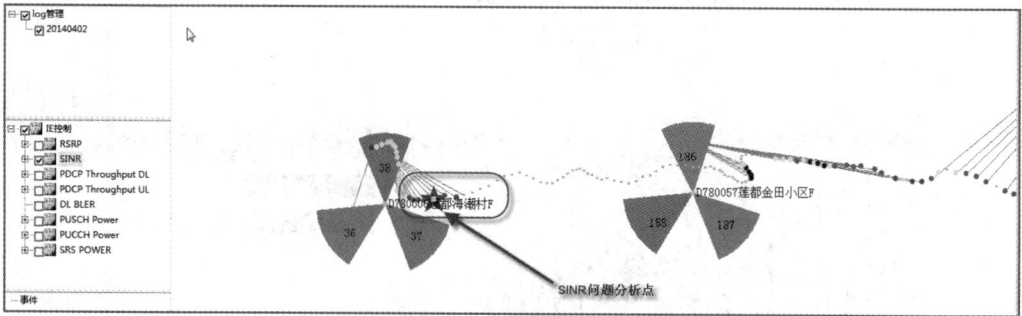

Fig. 4-3-8   Second screenshot of UE1 interference points

It is found that there is interference in this network, signal analysis is shown in Fig. 4-3-9.

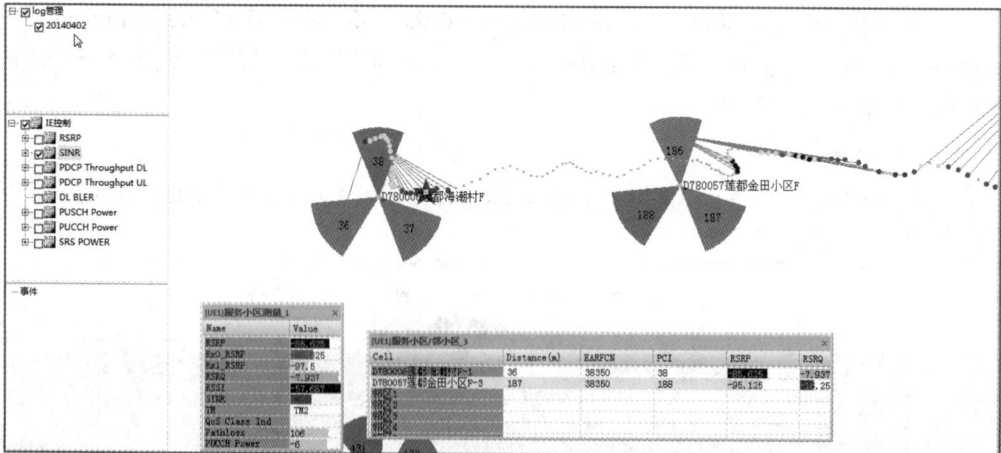

Fig. 4-3-9   Signal analysis

Through analysis, we can see that:

(1) Through an analysis of the following problem points, it can be seen that the problem point is covered by PCI 38, and the signal is very strong, but the SINR is poor.

(2) It is concluded by analysis that PCI 38 in the service area and PCI 188 in the neighbor

cell cover the road section together, but there is 2 left after PCI Mod 3, so the same PCI Mod results in poor SINR.

(3) It is suggested to swap PCI 188 and PCI 187, so as to improve the SINR value.

Task 4　High Level Signaling

This part mainly includes system message, random access, RRC connection, service establishment and release, and mobility management. In this course, we only give a preliminary introduction to the signaling function mechanism through examples, and the detailed signaling analysis is arranged at the technician level.

1. System information

The system information broadcasts to all UEs in the cell, with the aim to tell the public information of UE access layer and non-access layer, so that UE can understand the network configuration information before initiating the network connection. UE can obtain the following information through system information: system frame number, downlink system bandwidth, PHICH configuration information; cell selection and residence related information; resource configuration information of public channel; cell reselection information of same frequency and different frequency in the system; cell reselection information of different systems.

In the system information: MIB (Master Information Block), system frame number, downlink system bandwidth, and PHICH configuration information; SIB1, scheduling information and cell selection and residence related information of other SIB; SIB2, public channel resource configuration information; SIB3, reselection information (public parameter, used for cell reselection of the same frequency, different frequency and different systems); SIB4, same frequency cell reselection information; SIB5, reselection information of different frequency.

2. Random access

The function of random access is to obtain uplink synchronization with eNodeB and to apply for uplink resources. Application scenarios of random access are: UE in RRC_IDLE state initiates initial access; reestablishes connection after wireless link failure; handover process; UE in RRC CONNECTED state, uplink data arrives; UE in RRC_CONNECTED state, downlink data arrives; auxiliary positioning, network obtains time advance by using on-board access; classification of random access: random access based on competition and non-competition.

3. RRC connection

1) RRC connection established (successful)

(1) RRC connection request: carrying the UE's initial (NAS) identification and establishment reason, etc., which corresponds to the Msg3 of the random access process.

(2) RRC connection establishment: carrying the complete configuration information of SRB1, which corresponds to Msg4 of random access process.

(3) RRC connection is established: carrying the upstream NAS message (such as Attach Request message).

2) RRC connection establishment (failed)

Triggering reason: if eNodeB refuses to establish RRC connection for UE, a RRC connection rejection message is sent in reply.

3) RRC connection reconfiguration (successful)

Triggering reason: when it is necessary to initiate the management of wireless bearer, low level parameter configuration, handover execution and measurement control, this process is triggered.

RRC connection reconfiguration process: different configuration information can be carried according to different functions, and information units reflecting multiple functions can be carried in one message. Reconfiguration of RRC connection is completed: no actual information is carried, and only the RRC layer confirmation function is performed.

4) RRC connection reconfiguration (failed)

Triggering reason: if UE fails to perform the content in the reconfiguration message, UE goes back to the configuration before receiving the message and initiates the RRC connection re-establishment process.

5) RRC connection release

(1) Network release: this process is triggered when the network wants to release the RRC connection with UE.

(2) Local release: the RRC layer of UE actively releases the RRC connection following the instructions of the NAS layer. Instead of notifying the network side, it actively enters the IDLE state to restore the SRB1.

4. End to end business establishment/release process

The end to end business establishment/release process mainly includes Attach process, service request process, special bearer establishment process, special bearer modification process, special bearer release process, detach process, etc. The specific signaling content is arranged to be learned at technician level.

5. Mobility management

The mobility management mainly includes TAU and handover.

1) TAU

Tracking Area Update (TAU): when a mobile station moves from one TA to another, it must re-register its location on the new TA to inform the network to change the location information of the mobile station it stores. This process is TAU.

The update process of tracking area (TA) is always initiated by UE to inform the network that the current location of UE has changed.

Triggering scenarios of TAU: entering new TA; periodic update; entering E-UTRAN under URA-PCH state; entering E-UTRAN under GPRS READY state; load redistribution; RRC connection failure recovery; UE capability information change.

① The IDLE TAU process (including ACTIVE ID) is shown in Fig. 4-4-1. The TAI entered by UE in IDLE mode is not in the saved TAI list.

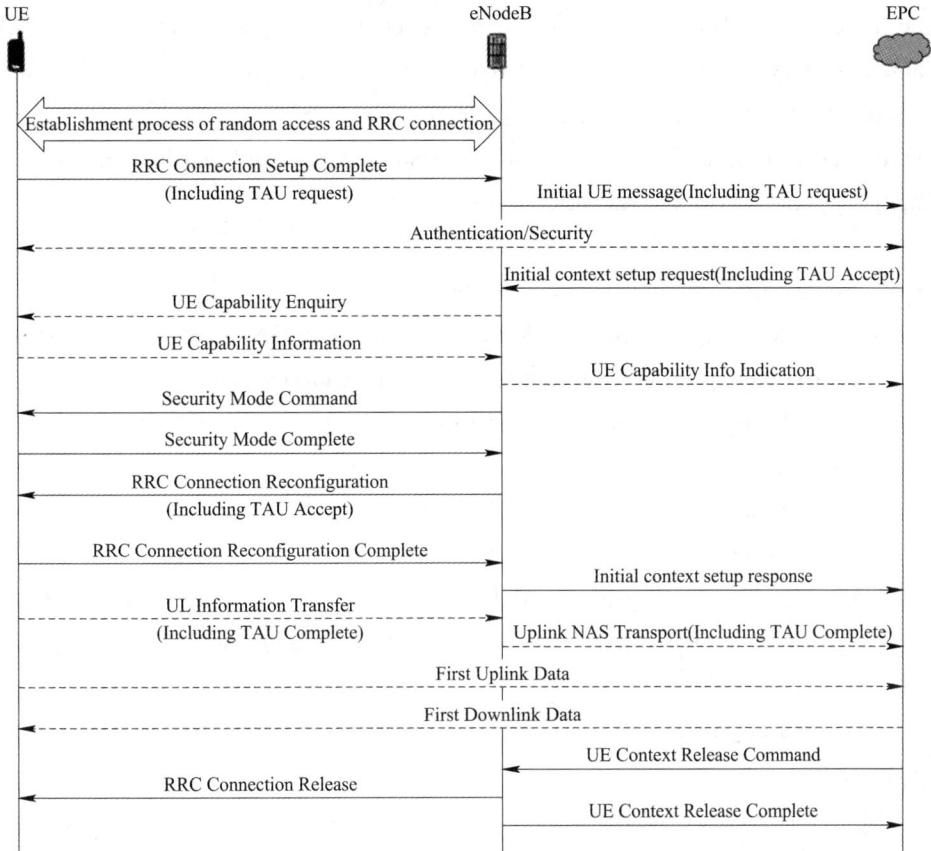

Fig. 4-4-1   IDLE TAU process

After going through the IDLE TAU process, the UE enters IDLE mode.

② The connected TAU process is shown in Fig. 4-4-2.

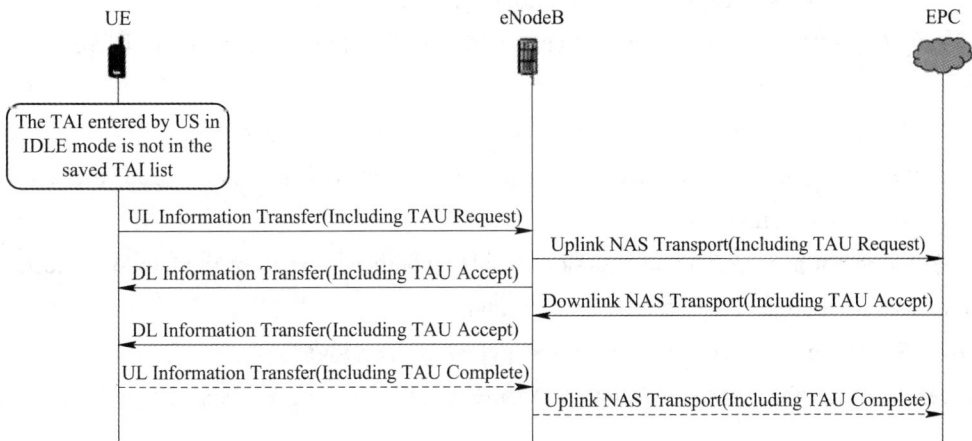

Fig. 4-4-2   Connected TAU process

2) Handover

Handover: in order to ensure the continuity of communication and the quality of service, the process of transferring the user's communication link to a new cell is called "handover".

The process is often triggered by wireless environment and network. ① Wireless environment: quality of received signal, level of received signal, distance between mobile station and base station, mobile station speed, etc. ② Network: congestion in service cell.

The overall process of handover includes measurement control distribution, measurement reporting, judgment process and execution process.

(1) Measurement control issued.

Below are measurement types that UE can perform.

① Same frequency measurement: Measure the downlink frequency point of the adjacent cell whose frequency point is the same as that of the current service cell.

② Different frequency measurement: Measure the downlink frequency point different from that of the current service cell.

③ System to system measurement with UTRA.

④ System to system measurement with GERAN.

⑤ Measurement between systems with CDMA 2000 lx EV-DO or CDMA 2000 1x RTT.

In RRC_IDLE state, the measurement parameter information of UE is obtained through E-UTRAN broadcast.

① SIB4: send out the measurement information of the same frequency neighbor cell (list of neighbor cell).

② SIB5: send out the measurement information of the neighbor cell of different frequencies (list of neighbor cell).

③ SIB6: issue the information of neighbor cell of UTRAN.

④ SIB7: issue the information of neighbor cell of GERAN.

⑤ SIB8: issue the information of neighbor cell of CDMA 2000.

In RRC_CONNECTED state, measurement configuration information is sent to UE through signaling.

(2) Reporting of measurement result. Triggering method includes event trigger and cycle trigger. Measurement event includes measurement event in the system and measurement eventsin different systems.

(3) Judgment process.The base station decides whether to perform the handover according to the measurement report reported by the terminal.

(4) Switch execution.There are three types of handover execution, eNodeB internal handover, X2 interface based handover (MME/ SGW unchanged), and S1 interface based handover.

Task 5   NG-RAN Overview

In 2019, ITU issued 5G standard, in 2020, 5G was officially put into commercial use, and 5G system was officially named IMT-2020. At present, the world's leading operating companies are accelerating 5G commercial deployment, and 5G industry is ready in the standards, products, terminals, security and commerce. Here is a preliminary introduction to the 5G network architecture and main features, and the key technologies will be arranged at the technician level.

1. From 4G to 5G

5G evolved from 4G, which is mainly reflected in all-round improvement. 5G makes up for the shortcomings of 4G technology and further improves the system performance in terms of throughput, delay, number of connections and energy consumption. 5G adopts digital all IP technology to support packet switching. It is neither a single technology evolution nor several new wireless access technologies, but integrates new wireless access technologies and existing wireless access technologies (2G, 3G, 4G, WLAN, etc.) to meet different needs by integrating a variety of technologies. It is a truly integrated network.

5G RAN is a 5G wireless access network, referred to as NG-RAN for short. It is an important part of 5G system, which has undergone great changes compared with 4G RAN. NG-RAN is composed of a group of gNBs (5G base stations), which are connected to 5GC (The 5th Generation Core Network) through NG interface. gNB consists of gNB CU (Centralized Unit) and one or more gNB DU (Distributed Unit). gNB CU and gNB DU are connected through F1 interface. The 5G RAN structure is shown in Fig. 4-5-1.

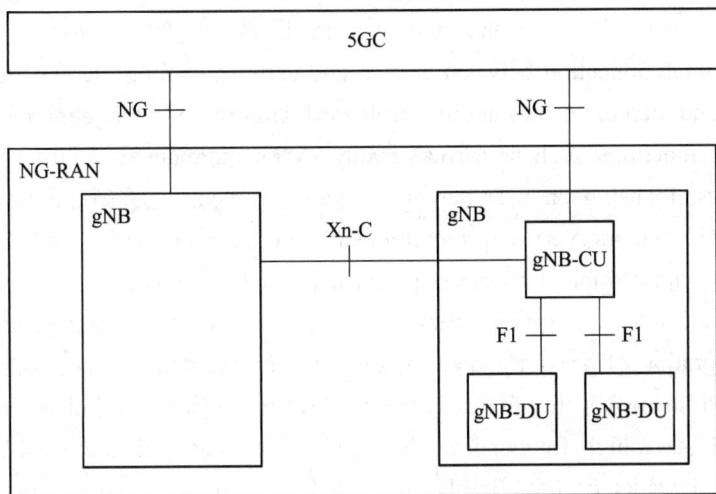

Fig. 4-5-1   5G RAN structure

5G communication technology achieves 1000 times more mobile data traffic per unit area than 4G. In terms of transmission rate, the data rate of typical users will be increased by 10~100 times, and the peak transmission rate can reach 10 Gb / s (100 Mb / s for 4G). 3GPP defines 5G applications into three scenarios: enhanced mobile broadband (eMBB, enhanced Mobile Broadband), massive machine type communication(mMTC, massive Machine Type Communication) and ultra reliable low delay communication(uRLLC, ultra Reliable Low Latency Communication). eMBB is mainly used for high-speed, large bandwidth mobile broadband services to improve the communication experience of personal services. Typical applications include ultra-high definition video, virtual reality, augmented reality, etc. The key to enhance the performance of mobile broadband is to maximize the bandwidth, achieve the ultimate traffic throughput, and reduce the delay as much as possible, so as to meet people's needs for super large traffic and high-speed transmission; mMTC is mainly for the IOT business of massive IOT links aimed at sensing and data acquisition. Typical applications include smart city, smart home, etc. Such applications require high connection density, and show industry diversity and differences at the same time; uRLLC is mainly for application scenarios with stable, reliable, real-time and controllable requirements. Typical applications include Internet of vehicles, telemedicine, industrial control, UAV(Unmanned Aerial Vehicle) control, etc.

2. From 5G to 6G

With the development of mobile communication technology, people's requirements for the transmission rate of mobile network are increasing. In the future, it will enter the 6G era. The goal of 6G system is to meet the needs of the information society after 2030. 6G system aims to realize intelligent communication, in-depth cognition, holographic experience and ubiquitous connection, and realize the seamless intelligent interconnection between man and everything. 6G network will use terahertz technology, that is, the frequency is between 0.1 THz~10 THz, which means wider spectrum resources and more access. At the same time, the transmission rate of 6G network will be more than 100 times that of 5G network, and the network delay will drop from the millisecond level of 5G to the microsecond level. The 6G network will meet the requirements of interconnection between heaven and earth, including the connection of ground equipment, low and medium orbit satellites, near earth aircraft and other equipment. Meanwhile, 6G will support functions such as virtual reality (VR), augmented reality (AR) and tactile Internet of things, focusing on user perception and experience. The reliability, security and privacy of 6G will become very key requirements. 6G is characterized by full coverage, full spectrum and full application. Full coverage means that 6G network will support the mutual communication and ultra dense connection among people, machines and goods, and develop towards the integration of heaven and earth, so as to achieve full coverage of all regions and fields. Full spectrum means that 6G system will develop towards millimeter wave, terahertz, visible light and other high frequencies. Full application means that the 6G system will be integrated into all walks of life, deeply integrated with artificial intelligence and big data technology, and have a revolutionary impact on the society.